Photovoltaic Partial Shading

This text comprehensively discusses the modeling of photovoltaic (PV) modules, PV array interconnections, multi-level inverters, distributed maximum power point tracking techniques, and static and dynamic PV array reconfiguration techniques. It gives a step-by-step procedure for hardware validation of the partial shading mitigation techniques.

This book:
- Focuses on the impacts and mitigation techniques related to partial shading problems associated with PV systems.
- Presents a step-by-step guide for addressing partial shading problems in PV systems.
- Covers methods like array reconfiguration through the Tom-Tom puzzle pattern and Arrow Sudoku pattern.
- Presents hardware validation of the partial shading mitigation techniques.
- Elaborates static and moving shading conditions in a detailed manner.

It will serve as an ideal reference text for graduate students and academic researchers in the fields of electrical engineering, electronics and communication engineering, environmental engineering, and renewable energy.

Photovoltaic Partial Shading
Principles and Methods

Anshul Agarwal, Anuradha Tomar, and
Venkata Madhava Ram Tatabhatla

CRC Press
Taylor & Francis Group
Boca Raton London New York

CRC Press is an imprint of the
Taylor & Francis Group, an **informa** business

Designed cover image: Volodimir Zozulinskyi/Shutterstock

First edition published 2025
by CRC Press
2385 NW Executive Center Drive, Suite 320, Boca Raton FL 33431

and by CRC Press
4 Park Square, Milton Park, Abingdon, Oxon, OX14 4RN

CRC Press is an imprint of Taylor & Francis Group, LLC

ISBN: 978-1-032-25953-6 (hbk)
ISBN: 978-1-032-25955-0 (pbk)
ISBN: 978-1-003-28583-0 (ebk)

DOI: 10.1201/9781003285830

Typeset in Nimbus Roman
by KnowledgeWorks Global Ltd.

Dedication

This book is dedicated to my family members and my acquaintances who has supported me in this work.

Contents

Foreword

Numerous variables, including tremendous capacity for meeting massive demand for electricity, widespread availability, coal depletion, and fights for sustainable development, have increased the need for power production from non-conventional sources of energy. Furthermore, in recent years, technology that opens the way to green power production has gained theoretical relevance. Solar energy has the most intense qualities of any renewable energy, which has prompted academics to develop in this new study arena. Solar energy has surpassed all other renewable energy sources due to its many advantages, including no carbon footprints, no maintenance, and no noise.

Although solar energy is good, flaws such as partial shadowing still exist. Partial shading (PS) is a phenomenon that happens in the Solar Photo-Voltaic Array (SPV) Array as a result of cloud passage, building shadow, and bird droppings, among other things. In practice, one of the most prevalent causes of power loss is shading of photovoltaic (PV) panels. PS is a critical topic for the research community to address in order to improve concerns in the area of solar array use. Shading shortens the life of a PV array, reduces its output power yield, and causes several peaks in its voltage-current and voltage-power characteristic curves. This book provides three distinct reconfiguration strategies to handle the aforementioned difficulties.

The initial technique of array reconfiguration is a hand-made mathematical puzzle pattern known as the Tom-Tom puzzle pattern, which is used to reconfigure the array's shaded and unshaded panels for shadow dispersion. The primary concept of a Tom-Tom problem is that the $N \times N$ grid should be filled with numbers ranging from 1 to N, with no number repeating in any row or column. The sub-grids are also included in the $N \times N$ grid. On the left-top corner of each sub-grid is a number and the operation to be done on the numbers of the relevant grid. A sub-number of grids should be filled in such a way that the specified number may be obtained by utilizing these numbers and the mathematical operation specified on them. Different shadings are evaluated to verify the suggested reconfiguration process. The suggested descriptor functioned well in all shading circumstances.

The second strategy, which is known as the Arrow Sudoku pattern, involves spreading the shade along the same column. The main difference between the original Sudoku and the variant known as Arrow Sudoku is that the latter makes use of sub-grids in addition to arrows. Both games are otherwise quite similar. One end of the arrow (the tail end) comprises of sub-grids, each of which has numbers somewhere inside it. The head end of the arrow points in the direction of a number that is encircled in the illustration. The Arrow Sudoku problem may be solved by using the same rules as the traditional Sudoku puzzle; however, the total of the numbers in the sub-grids of the arrow's tail end must be equal to the number that is ringed in the arrow's head. To solve an Arrow Sudoku, you will need more than simply the logic of Sudoku. This is because the puzzle also contains arrows. The Arrow Sudoku pattern serves as the organizing principle for the panel layout at the physical level. The fact that the suggested method performs better than the current state-of-the-art methods in every instance of the PS situations offers clear evidence in favor of the usefulness of the proposed method.

Finally, an experimental setup was created to test the performance of all suggested and traditional setups under realistic shading circumstances. The various irradiance levels are simulated (artificial uniform and PS patterns) by employing different hues. When compared to several state-of-the-art methodologies, all of the offered approaches produced promising outcomes for all sorts of shadowing circumstances.

Author Biography

Dr. Anshul Agarwal graduated from the Uttar Pradesh Technical University, Lucknow, India with a B. Tech degree in Electrical and Electronics Engineering in 2007 and completed his M.Tech (Gold Medalist) in Power Electronics and ASIC Design in July 2009, and Ph.D. in 2013 in MNNIT, Allahabad. He is currently the faculty in the Electrical Engineering Department at Babasaheb Bhimrao Ambedkar University (BBAU) Lucknow. He served as an assistant professor at the National Institute of Technology Delhi (NIT Delhi) from December 2013 to July 2023. Before this, he served as a lecturer at the National Institute of Technology Hamirpur from February 2012 to December 2013. His research interests include power electronic devices, modeling of converters/inverters, AC to AC power conversion, renewable energy, integration of grids, hybrid vehicles, FPGA-based converters design, and solar and wind energy integration. Dr. Anshul Agarwal has published more than 60 papers in reputed International Journals, and more than 65 papers in reputed IEEE International Conferences. He has also published 2 books and 11 Book chapters in different Books. He has guided 6 Ph.D. scholars and currently 2 Ph.D. theses are submitted under his supervision. He has been awarded by POSOCO Power System (PPSA-2014) for top-level Ph.D. research work. He has also served as a guest editor in special issues for various journals. He is an active reviewer of various international conferences and reputed journals. He has more than 12 years of research and teaching experience.

Dr. Anuradha Tomar has 12 years plus experience in research and academics. She is currently working as an assistant professor in the Instrumentation & Control Engineering Division of Netaji Subhas University, Delhi, India. Dr. Tomar has completed her Postdoctoral research in Electrical Energy Systems Group, from Eindhoven University of Technology (TU/e), the Netherlands, and has completed European Commission's Horizon 2020, UNITED GRID, and UNICORN TKI Urban Research projects. She received her B.E Degree in Electronics Instrumentation & Control with Honours in the year 2007 from the University of Rajasthan, India. In the year 2009, she completed her M.Tech Degree with Honours in Power Systems from the National Institute of Technology Hamirpur. She received her Ph.D in Electrical Engineering, from the Indian Institute of Technology Delhi (IITD), India. Dr. Anuradha Tomar has committed her research work efforts toward the development of sustainable, energy-efficient solutions for the empowerment of society, and humankind. Her areas of research interest are the operation & Control of microgrids, photovoltaic systems, renewable energy-based rural electrification, congestion management in LV distribution systems, artificial intelligence & machine learning applications in power systems, energy conservation, and automation. She has authored or co-authored 69 research/review papers in various reputed International, National Journals, and Conferences. She is an Editor for books with international publications like Springer and Elsevier. She has four granted Patents in her name and five filed Indian patents in her name. Dr. Tomar is a senior member of IEEE and a life member of ISTE, IETE, IEI, and IAENG.

Dr. Venkata Madhava Ram Tatabhatla was born in Warangal, India, in 1992. He received the B.Tech. and M.Tech. degrees in electrical engineering and power electronics engineering from Kakatiya University, Warangal, India, in 2013 and 2015, respectively. He achieved his doctoral degree with the Department of Electrical and Electronics Engineering, NIT Delhi, India in 2021. He is currently working as an assistant professor in the Electrical and Electronics Engineering Department at B. M. S. College of Engineering, Bengaluru. He has authored more than ten research articles in reputed journals and conferences. His research interests include power electronics, renewable energy, applications of power electronics in renewable energy, design and implementation of solar PV systems, renewable energy resources, power management for hybrid energy systems, and smart grids. He is an active reviewer of various international conferences and reputed journals.

1 Introduction

1.1 OVERVIEW

The consumption of electric power around the globe is increasing gradually every day. In guise of the gradual increment in power consumption, the urge of energy production from power plants rises too. The energies like coal, fossil fuels, oil, and gas are known as conventional energies, which are exhaustible in nature [1–3]. The conventional energies lead to pollution and environmental deterioration as they emit greenhouse gases, smoke, and ash [4]. Hence, there is a rise in the requirement for utilization of renewable energies like solar, biogas, wind, and geothermal, etc [5–7]. Out of all renewable energies, solar energy has intense characteristics that stimulated researchers to evolve in this new research domain [8–13]. In solar energy the unlimited energy of the sun is harnessed using solar panels which convert the solar irradiance into electricity. The cost of per kW power produced by solar energy is cheaper compared to power generated by fossil fuel-based power plants. Solar power plant also requires less maintenance compared to other forms of energy generation methods.

Although the advantages are many which makes solar energy one of the most preferred sources of renewable energy, there are few major disadvantages which hinders its power generation capabilities [14, 15]. One of the main drawbacks of solar energy is its dependence on the environmental conditions like temperature and irradiance level. Consequently, the above-mentioned factors are providing hindrance between the real energy produced and the energy estimated. Keeping aside the temperature which under most cases remains constant throughout the solar array, then the only major goal is to have uniformity in the level of solar irradiance. But to achieve uniform irradiance level is not easy as different conditions like moving clouds, accumulation of dust and dirt on the solar panel, shadow from trees and buildings, etc causes partial shading of the solar panels. When the solar panels connected in an array receives different levels of irradiance then the panels which are getting lesser irradiance are termed as partial shaded.

The partial shading condition causes the reduction in the power generated by the solar panels. Although the partially shaded panels generated less current compared to other uniformly irradiated panels but due to series connection they are forced to carry higher amount of current which leads to reverse biasing of the shaded panels and they start acting as load. This phenomenon leads to the increment of temperature of the shaded panels and causes Hot-Spot problem which damages the panels permanently. To overcome this problem, bypass diodes are used in parallel with each solar panel. Whenever the panels get reverse-biased, the bypass diodes get activated and as per the name provides bypass path for the current to flow without damaging the panels. As every solution comes with some added difficulties same is the case with bypass diodes which provides a solution for avoiding Hot-Spot phenomenon but creates a

DOI: 10.1201/9781003285830-1

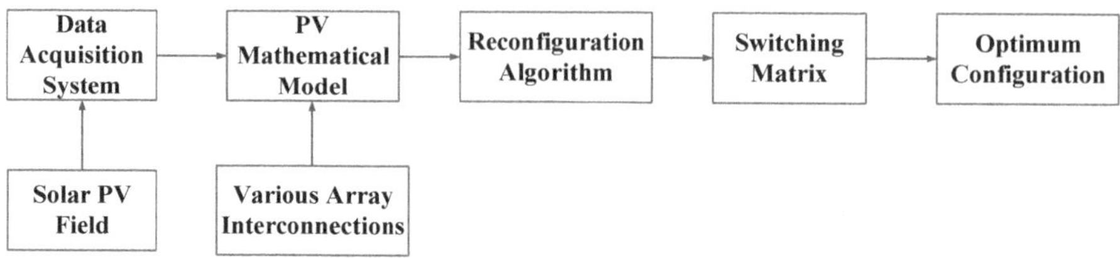

Figure 1.1   Dynamic photo-voltaic array reconfiguration (DPVAR) scheme.

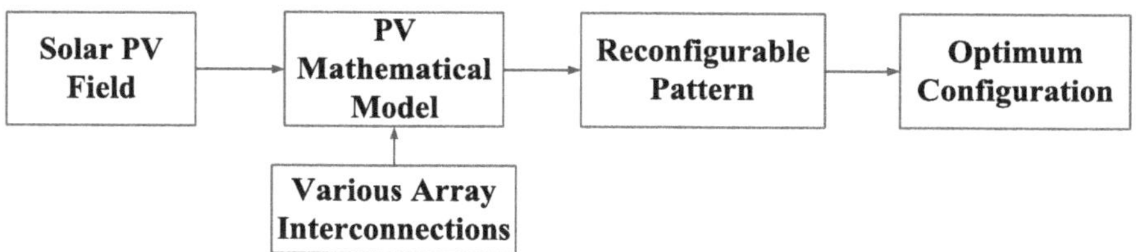

Figure 1.2   Static photo-voltaic array reconfiguration (SPVAR) scheme.

problem of multiple peaks in the P-V curve. The presence of multiple peaks in the P-V curve makes the maximum power point tracking difficult. This results in decrement in the actual output power compared with the rated power output. This also effects the efficiency of the solar panels which is already not very high and hence the overall returns over the investment of the solar farms are not justified.

Based on the literature, these are predominantly divided into (1) Static Photo-Voltaic Array reconfiguration (SPVAR) strategies and (2) Dynamic Photo-Voltaic Array reconfiguration (DPVAR) strategies. Each have its own advantages and disadvantages which are discussed in coming sections.

In dynamic reconfiguration, the electrical connections between the panels in a PV array changes dynamically with respect to irradiance levels so as to enhance maximum power. Figure 1.1 represents the control structure of DPVAR. It comprises of a data acquisition system (DAS) that needs appropriate measurement equipment to gather field information for the PV model [16–18]. While, the unresolved parameters are evaluated by the mathematical model and then furnishing them to the reconfiguration algorithm. Here algorithm discovers optimal arrangement and then correspondingly control algorithm is provided to the switching matrix. However, in SPVAR, the physical locations of shaded and unshaded panels are changed in this technique, without altering their electrical connections to enhance power output by shade dispersion. Figure 1.2 illustrates the different stages of SPVAR approach. The reconfiguration schemes of PV panels are necessary for shade dispersion of diluting the concentrated shade.

The above said reconfiguration approaches can be implemented on a variety of applications.

- If the shadow projected by a steady, fixed object affects the PV plant: This case is very usual for PV plants which are integrated into buildings or arranged on roofs. For instance, in [19], DPVAR strategies have been practiced for building integrated PV systems (BIPVS). The life cycle and the sustainable assessment of a BIPVS module have been extensively studied. BIPVS are given substantial attention as they allow to grab the benefit of wide-area revealed to sunlight. In these kinds of applications, the SPVAR strategies can also be used to distribute the concentrated shade on to the entire array so as to mitigate the effect of shading.

- Effect of passing clouds on the part of a huge PV plant: Generally, in this instance, distributed drop in irradiance exists above the PV array. The level of irradiations in this case depends on the fastness of the moving clouds which produce different shading levels on the PV array and they change suddenly paving the way to a huge degradation in efficiency of the PV array. This shade may be present for a longer duration time, which in turn is responsible for a depletion of the energy produced by the source. Eventually, if a failure occurs for one or more solar modules, automatic disconnection happens on employing dynamic approach [20]. Nevertheless, with regard to static technique, rearrangement of PV modules is in such a way that the dispersion of shade is equal.

1.2 MODELING OF SOLAR PHOTO-VOLTAIC CELL

In order to understand the characteristics and execution of any solar plant, the main criteria are to have an understanding about the basic building blocks of the solar plant which are solar cells. The connection of various solar cells leads to the formation of solar panels which are further connected in various interconnection fashions to form solar array. The modeling of the solar cells helps to gain insight about the working of the solar cells. The non-linear characteristics of the solar cells makes modeling a difficult task to achieve the required accuracy and optimal design. The three common models used are single-diode PV cell model, two-diode PV cell model, and three-diode PV cell model. Out of these, single-diode PV cell model is largely used because it is simple to understand and implement. The ideal, practical, and simplified equivalent circuit model of PV cell is shown in Fig. 1.3(a).

The current generated by the solar cell is given by I_{solar}. The diode connected in the circuit produces the reverse current I_{diode} and the circuit has two resistors R_{shunt} which is connected in shunt to the diode and R_{series} which is connected in series to the output terminal. R_{series} is used to determine slope near open circuit condition and R_{shunt} is used to determine slope near short circuit condition in I-V curve.

Applying Kirchoff's Current Law (KCL) at outer node, the output current I is given by

$$I = I_{solar} - I_{diode} - I_{shunt} \tag{1.1}$$

The shunt current I_{shunt} is given as

$$I_{shunt} = \frac{V + IR_{series}}{R_{shunt}} \tag{1.2}$$

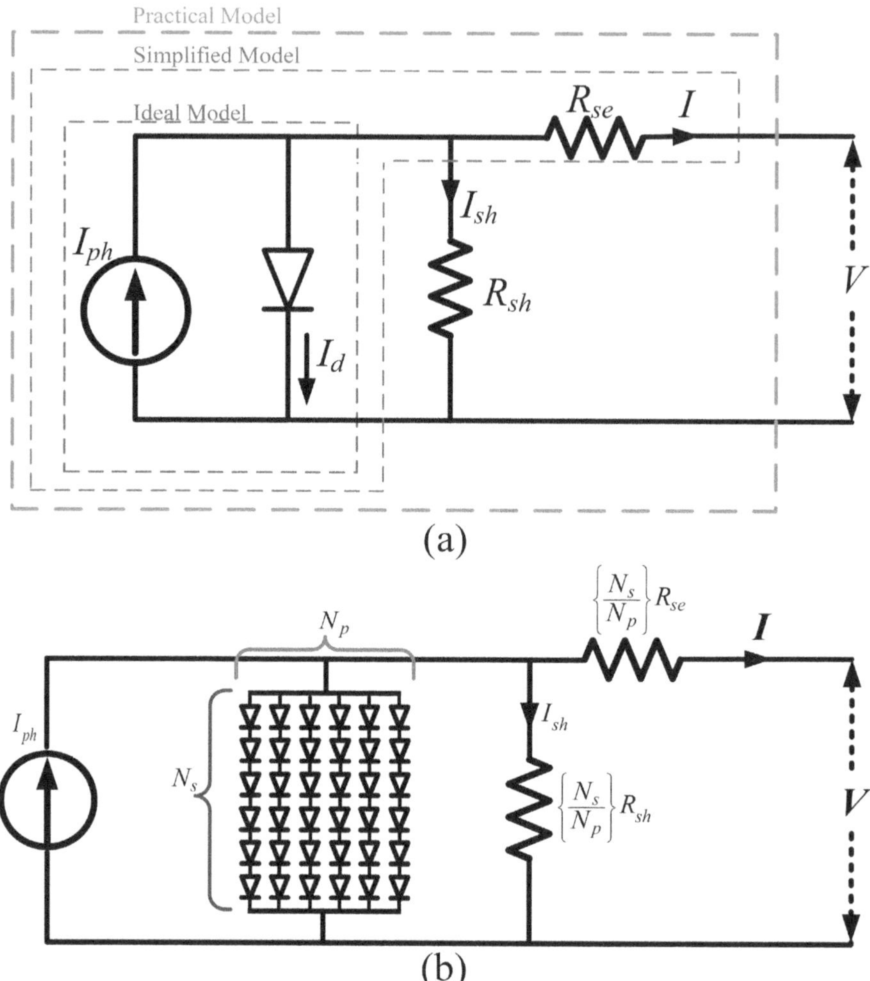

Figure 1.3 (a) SPV cell equivalent circuit model (b) SPV module equivalent circuit.

The diode current I_{diode} is given as

$$I_d = I_s \left[\exp\left(\frac{V + IR_{se}}{nV_t} \right) - 1 \right] \tag{1.3}$$

Substituting eqs. (1.3) and (1.2) in (1.1) gives

$$I = I_{ph} - I_s \left[\exp\left(\frac{V + IR_{se}}{nV_t} \right) - 1 \right] - \left(\frac{V + IR_{se}}{R_{sh}} \right) \tag{1.4}$$

From equations, it is evident that the current is inversely proportional to the temperature and power is inversely proportional to the temperature. Therefore, when the

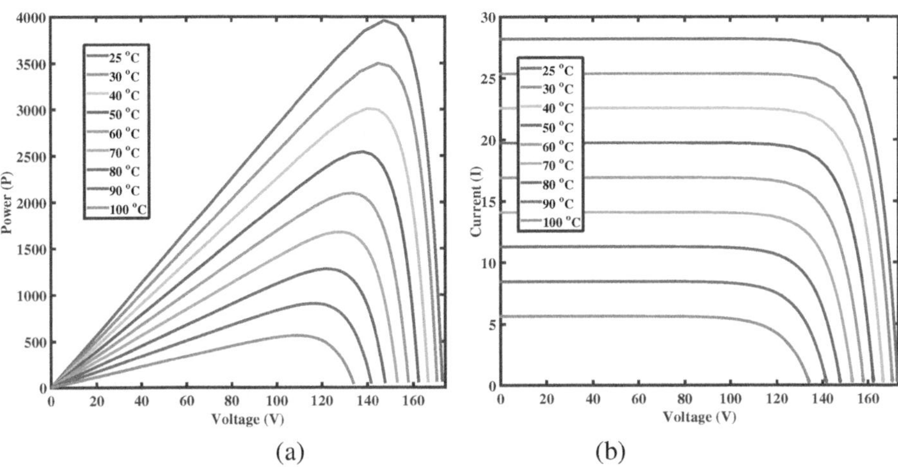

Figure 1.4 (a) Voltage-Power characteristics (b) SPV array Voltage-Current characteristics at variable temperature levels with a constant irradiation of $1000\text{W}/\text{m}^2$.

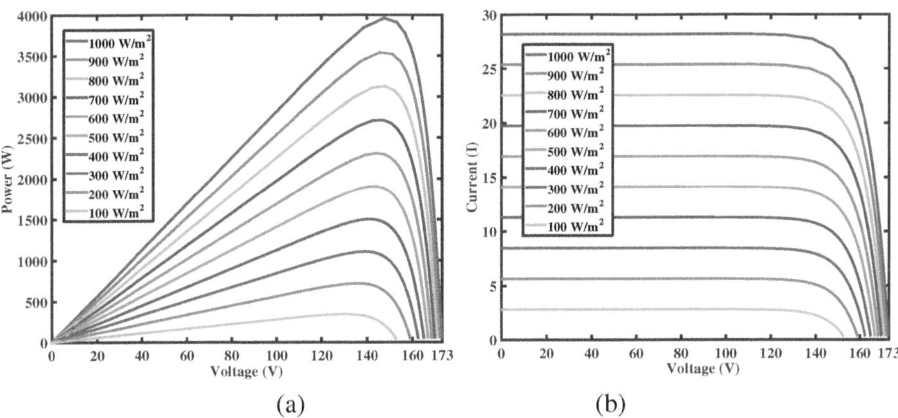

Figure 1.5 (a) Voltage-Power characteristics (b) SPV array Voltage-Current characteristics at various temperature levels at a constant temperature of $25^\circ C$.

temperature of the PV module is increased beyond Standard Test Conditions (STC), the current and power decreases as depicted in Figs. 1.4(a) and 1.4(b). On any given day, the irradiations incident on the solar panel continuously varies. This results in the change in power as shown in Figs. 1.5(a) and 1.5(b).

1.3 PARTIAL SHADING AND MISMATCH EFFECTS

A PV cell alone can produce a very minimal power output of 1 to 1.5 W which in most cases is not sufficient to fulfill the load-side demand. In order to extract

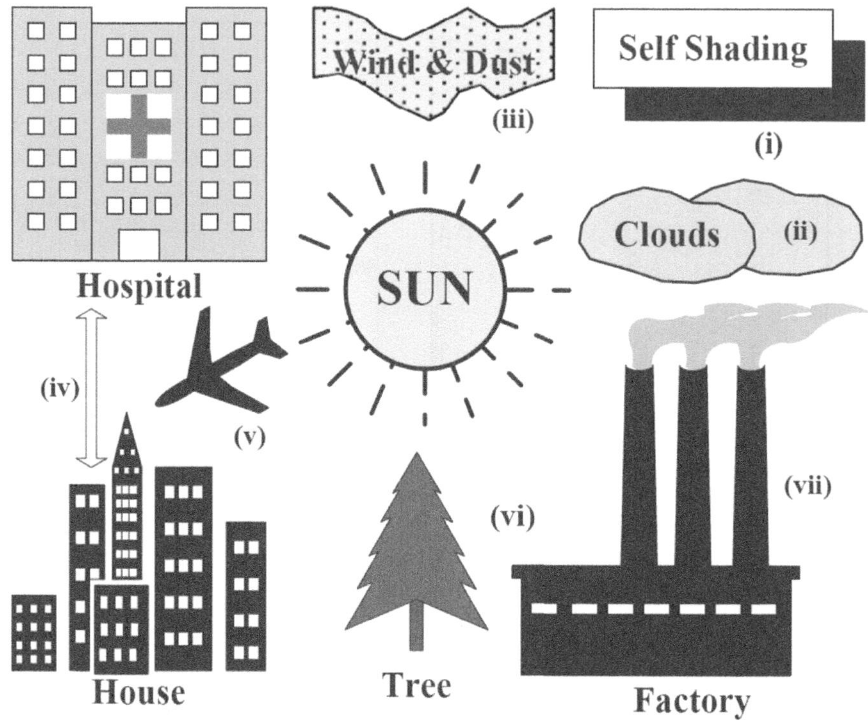

Figure 1.6 Causes of Partial shading conditions [29].

useful level of power, PV cells are connected in series to form PV panels. Now as per the load requirement these PV panels or modules are further connected in series and parallel format so that they can achieve the required current and voltage level. This overall connection of PV panels in series and parallel forms PV array. As PV array covers large surface area so that they are susceptible to non-uniform irradiance condition caused by clouds, dust and dirt, the shadow of buildings and trees as shown in Fig. 1.6 [21–23]. This condition in which PV panels connected in PV array are not subjected to same irradiance level is called as partial shading.

The PV panels which are exposed to lesser solar irradiance tend to generate low level of current compared to PV panels which are fully irradiated and hence this leads to the problem of current imbalance between the PV panels connected in the same series connection [24–26]. The partially shaded PV panels are compelled to carry higher amount of current which causes them to get reverse biased and start acting as load to the existing system. The overall power output decreases and also causes increase in the temperature of the partially shaded panels [27, 28]. The rise in temperature leaves Hot-Spot which damages the PV panels. The Voltage-Current and Voltage-Power characteristic curves under normal (STC) and partial shading conditions are given in Figs. 1.7(a) and 1.7(b), respectively.

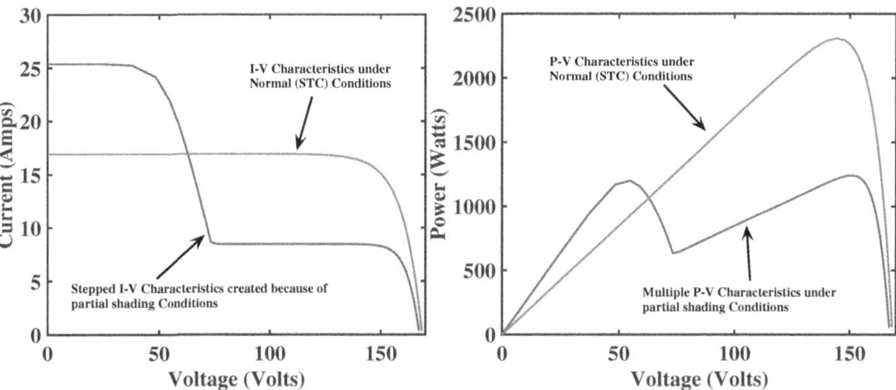

Figure 1.7 Voltage-Current and Voltage-Power characteristic curves under normal and partial shading conditions.

1.4 PARTIAL SHADING REDUCTION TECHNIQUES

For minimizing the partial shading (PS) losses, various mitigation techniques have been described in the literature which are predominantly divided into two categories, Passive and active mitigation techniques.

1.4.1 MITIGATION OF PARTIAL SHADING WITH PASSIVE TECHNIQUES

Under this technique, a diode is connected in parallel across each PV panel. The diodes are termed as bypass diodes because they provide an alternative path for the current to flow bypassing the shaded PV panel. It can be observed that bypass diodes is needed to control the problem of Hot-Spot caused by partial shading but at the same time bring in another problem of multiple peaks in the P-V curve which makes the MPPT control difficult. So apart from bypass diodes, mobile mitigation techniques also need to be applied to avoid the multiple peaks in the output characteristics. In this approach, bypass diode is used as passive element in common for minimizing partial shading effects. Voltage-Current characteristic curve displays multiple steps and Voltage-Power characteristic curves display multiple peaks because of the insertion of bypass diodes, which are depicted in Figure 1.7.

1.4.1.1 Interconnection Topologies

The power generated by a single solar panel will be low. To enhance the electrical specifications and to meet the load demand, these panels are interconnected. This interconnection of panels is generally called as configuration. In literature, different topologies such as Series (S), Parallel (P), Series-Parallel (SP), Total Cross Tied (TCT), Bridge-Link (BL) and Honey Comb (HC) have been explained [30, 31]. The arrangement of these topologies is depicted in Fig. 1.8. In series configuration (Fig. 1.8(a)), the desired output voltage increases as all PV panels are connected in series

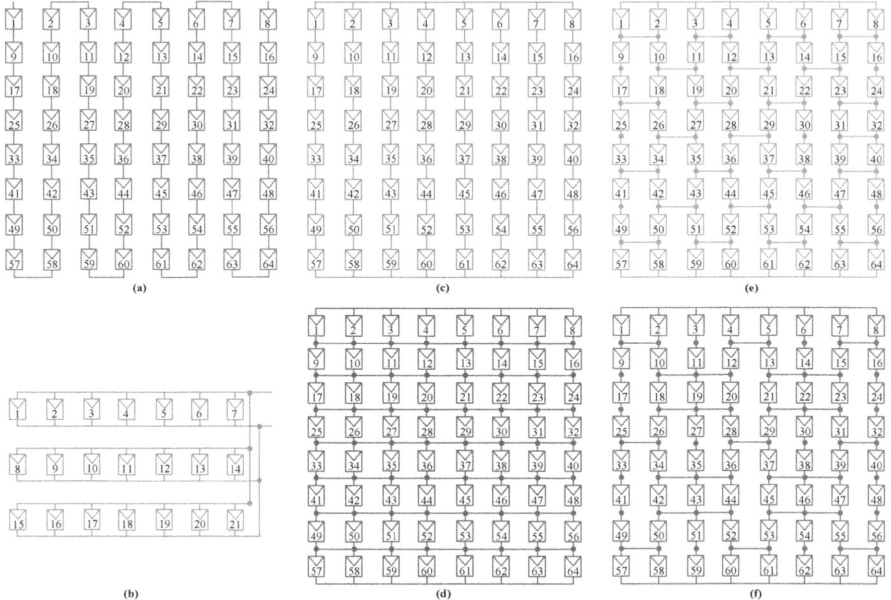

Figure 1.8 Different PV configurations (a) Series (S) (b) Parallel (P) (c) Series-Parallel (SP) (d) Total Cross Tied (TCT) (e) Bridge Link (BL) (f) Honey Comb (HC).

whereas the output current is equal to the panel current [32, 33]. In parallel configuration (Fig. 1.8(b)), the output voltage will be equal to panel voltage and the output current increases. Parallel configuration is advantageous over series because the maximum power obtained when connected in parallel configuration is greater than that obtained when connected in series.

The series connected panels are non-resilient to shade whereas the parallel connected panels are resilient. So, the number of panels are interconnected in a matrix arrangement with rows in series and columns in parallel known as a series-parallel (SP) configuration depicted in Fig. 1.8 (c). These boost both output voltage as well as current at the same time. The connection of cross ties across each junction of series-parallel configuration forms a Total Cross Tied configuration as shown in Fig. 1.8 (d). Thereafter, two different configurations viz., Bridge Link (BL) and Honey Comb (HC) came into existence to reduce the wiring losses of TCT configuration in [34]. BL is inspired by the Wheatstone bridge and bridge rectifier form. Almost half of its connections are refrained and hence the line losses can be drastically decreased (Fig. 1.8 (e)) and HC is honeycomb like architecture (hexagon shape) as shown in Fig. 1.8 (f) of the SPV Array and is formed by the combination of TCT and BL configurations. Under uniform conditions, SP, BL, HC, and TCT connected SPV arrays generate same amount of output power [35]. Table 1.1 gives the qualitative comparison of Photo-Voltaic Interconnection Topologies.

Table 1.1

Qualitative Comparison of Photo-Voltaic Interconnection Topologies

Parameters	Photo-Voltaic Interconnection Topologies			
	SP	BL	HC	TCT
Number of Cross Ties	Nil	Low	Moderate	High
Redundancy Level	Low	Moderate	Moderate	High
Reliability	Low	Moderate	Moderate	High
Complexity in Connection	Low	Moderate	Moderate	High
Real-Time Applications	Widely Used	Less Used	Less Used	Widely Used
Power Output under normal conditions	Equal	Equal	Equal	Equal
Power Output under shading conditions	Low	Moderate	Moderate	High
Cost	Moderate	High	High	Moderate

1.4.2 MITIGATION OF PARTIAL SHADING WITH ACTIVE TECHNIQUES

In order to decrease the PS effects, the active techniques are more efficient than passive techniques. These are divided into three categories viz., Distributed MPPT strategies, Multi-level inverters, and reconfiguration methods.

1.4.2.1 Distributed MPPT Strategies

Every panel or group of panels, in this technique, has its individual MPPT, hence preventing losses. Further, this technique avoids the establishment of bypass diodes. Thus, the equivalent losses are ignored. Many inverters and converters are needed for each panel. Furthermore, an extremely complex structure is required as depicted in Fig. 1.9.

1.4.2.2 Multi-Level Inverters

Multi-level inverters (MLIs)-based MPPT techniques are a new inclusion to the PV panels for decreasing losses obtained by partial shading. MLI topologies like cascaded H-bridge, flyback capacitor, and diode-clamped are employed to decrease the shading effects on a panel or array by monitoring output voltages. These also decrease AC voltage harmonics and voltage stress. According to this approach, every

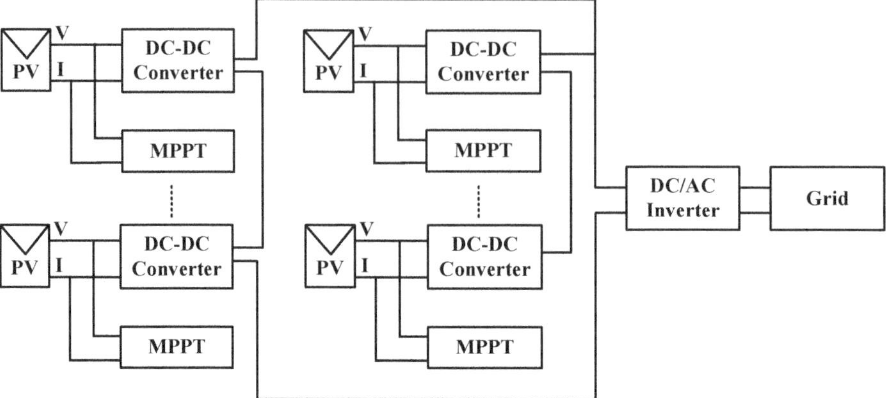

Figure 1.9 Design of PV panels with MPPT strategies.

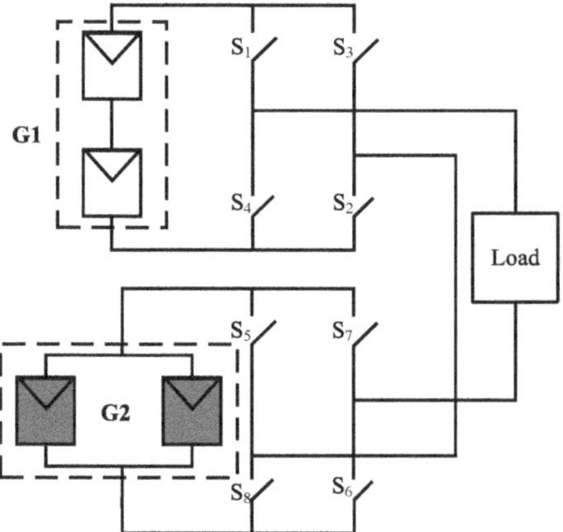

Figure 1.10 Group of PV panels with multi-level inverters (MLIs)-based MPPT techniques.

panel or a group of panels has individual inverters (Fig. 1.10), which in turn increments the price of the system. Another important active technique to improve the output parameters under partial shading conditions is extensively explained in the next sub-section "Reconfiguration Strategies."

1.4.2.3 Reconfiguration Methods

Under PSCs, reconfiguration techniques are the most assuring ones to enhance output power. Here the PV panels are reconfigured based on different levels of irradiation. As stated earlier, these are predominantly divided into (1) Static Photo-Voltaic Array

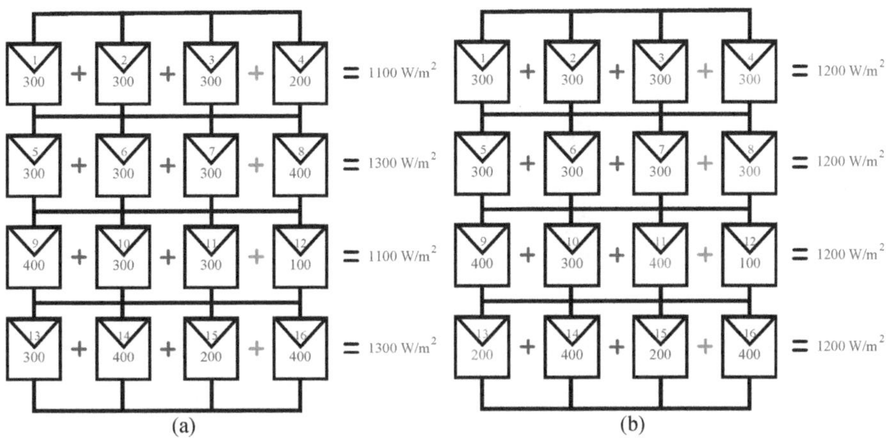

Figure 1.11   Example for dynamic reconfiguration using irradiance equalization in TCT arrangement (a) Before reconfiguration: the sum of irradiance levels for each row is 1100W/m^2, 1300W/m^2, 1100W/m^2 and 1300W/m^2 (b) After reconfiguration: the panels 4 and 13, 8 and 11 are interchanged and the resultant irradiance is 1200W/m^2.

reconfiguration (SPVAR) strategies and (2) Dynamic Photo-Voltaic Array reconfiguration (DPVAR) strategies.

1.4.3   DYNAMIC PHOTO-VOLTAIC ARRAY RECONFIGURATION (DPVAR) STRATEGIES

In order to mitigate the shading effects, in DPVAR strategies, the PV module connections in this approach change dynamically. The real dynamic reconfiguration technique chooses the optimum PV array that has the capacity to curtail mismatch effects.

[36, 37] presents an optimization algorithm based on equalization index (EI). The fundamental objective is to stabilize the irradiance level of panels in a row using electrical and electronic equipment, such as sensors and switches, to reduce mismatch losses. The switching between the panels in an SPV array is regulated by this algorithm. The level of irradiance in each row is denoted by G_i and is computed as

$$G_i = \frac{\sum_{q=1}^{m} G_{pq}}{m} \tag{1.5}$$

An example for this method is depicted in Fig. 1.11. This algorithm uses the following expression to calculate the equalization index for each reconfiguration. This algorithm chooses an optimum reconfiguration based on minimum EI and switching operations.

$$EI = Max(G_i) - Min(G_i) \; \forall \, i \tag{1.6}$$

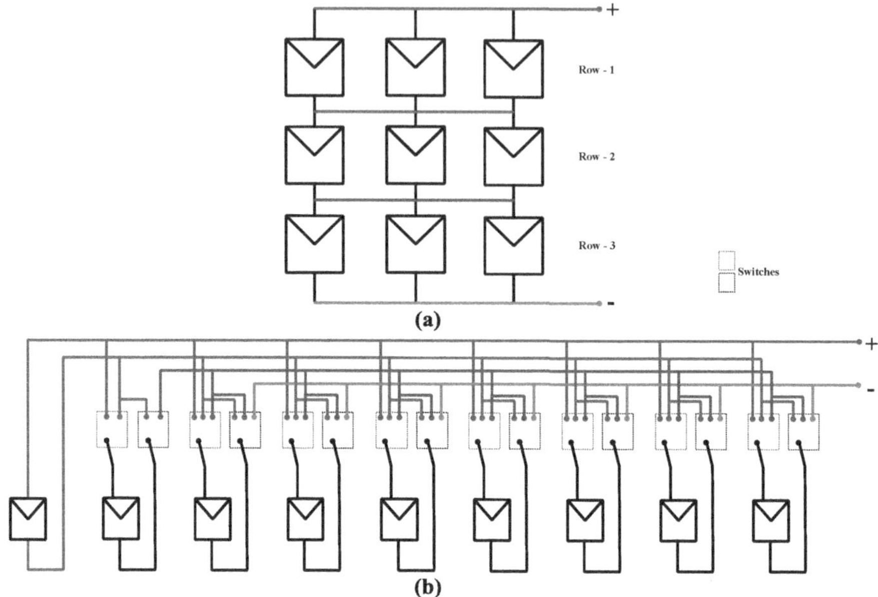

Figure 1.12 (a) TCT connected 3×3 SPV Array (b) Reconfigured Switching Matrix used for 3×3 SPV Array.

A PV generator using EAR is proposed in [38, 39]. This consists of static and reconfigurable parts. Static part has to reach inverter constraints and IE algorithm controls reconfigurable part. Let p be the number of rows and q be the number of columns in an array, the number of reconfigurations are determined by $(p \cdot q)!$. Let "C" be the total reconfigurations obtained with different output power, which is given as

$$C = \frac{(p \cdot q)!}{p! \cdot ((q)!)^p} \tag{1.7}$$

The maximum possible reconfigurations are achieved by using electrical and electronic equipment such as switches, sensors and control algorithms. This switching matrix method is developed in Fig. 1.12 for 3×3 SPV Array. The number of switches (N_s) used for this reconfiguration is given by

$$N_s = 2N \tag{1.8}$$

Authors have developed an iterative and hierarchical sorting algorithm to identify the best reconfiguration with IE principle in [40]. Figure 1.13 clearly depicts an illustration of this proposed algorithm. In this algorithm, the switches used are single pole single throw switches. According to the authors, the number of switches used for reconfiguration through this algorithm is $N_s = N_{pv}(p^2 - 1)$. All the panels in this algorithm are arranged in descending order of their level of irradiation forming a

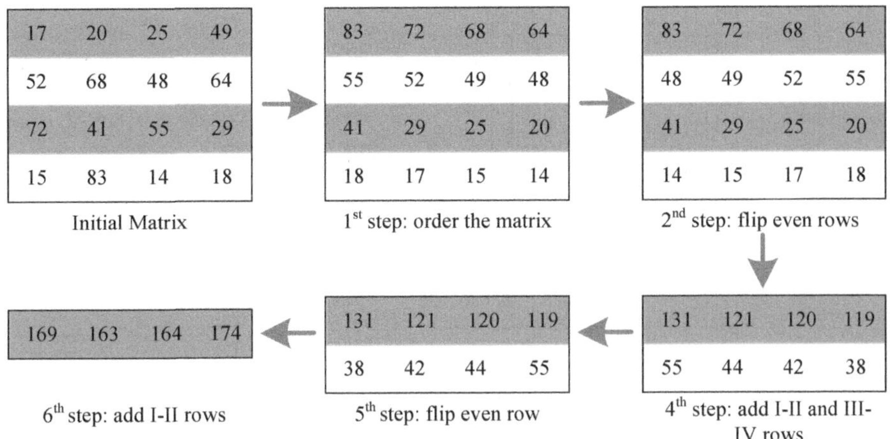

Figure 1.13 steps (labels within figure):
Initial Matrix
1st step: order the matrix
2nd step: flip even rows
6th step: add I-II rows
5th step: flip even row
4th step: add I-II and III-IV rows

Figure 1.13 Illustration for sorting algorithm.

matrix in the first step. All the even rows are tossed from left to right and added to the previous odd rows producing a row matrix in its subsequent steps.

A flexible reconfiguration structure is introduced with fixed solar cells in [41]. Figure 1.14 depicts the switching matrix that joins the fixed and adaptive solar cells using a control algorithm to maintain equal row current. The fixed part and adaptive parts are connected through a matrix called as switching matrix and when the shade occurs, the switching matrix automatically adjusts to maintain equal row current and under no shade conditions, the connections will not be changed.

In order to make up for the irradiance drop, the most irradiated solar cell from adaptive part is connected to the fixed part. The switching between these parts (switching matrix) is controlled by model based and bubble sort algorithms. Subsequently, [42] proposed adaptive reconfiguration method based upon a neural network. The output power of the solar PV array is estimated by taking the solar irradiance, sun position angles and ambient temperature as input to train multilayer feed-forward neural network.

A dynamic reconfiguration algorithm is proposed in [25] in order to diminish the power fluctuations that are caused by spatial irradiance profiles under practical scenarios. Spatial Irradiance, which usually arises in real-time PV systems, has shading scenarios caused by moving cloud irradiation profile. Authors have taken irradiation profiles by taking sky images and incident shading patterns. This proposed algorithm is based on Irradiance Equalization (IE) and process of this algorithm is as follows: A 4×4 SPV array with different levels of irradiations is considered. The incident Irradiation should be in the order of $G_1 < G_2 < G_3 \ldots < G_{16}$ as depicted in Fig. 1.15(a). In the next step, calculate the IE for each of the SPV array. Based on the IE index calculated for each row, the algorithm will decide to reconfigure the panels in different rows so as to maintain same row irradiations. The average value of irradiance lies at 801.56W/m^2. Row-1 and Row-4 exceeds the average values and it is necessary

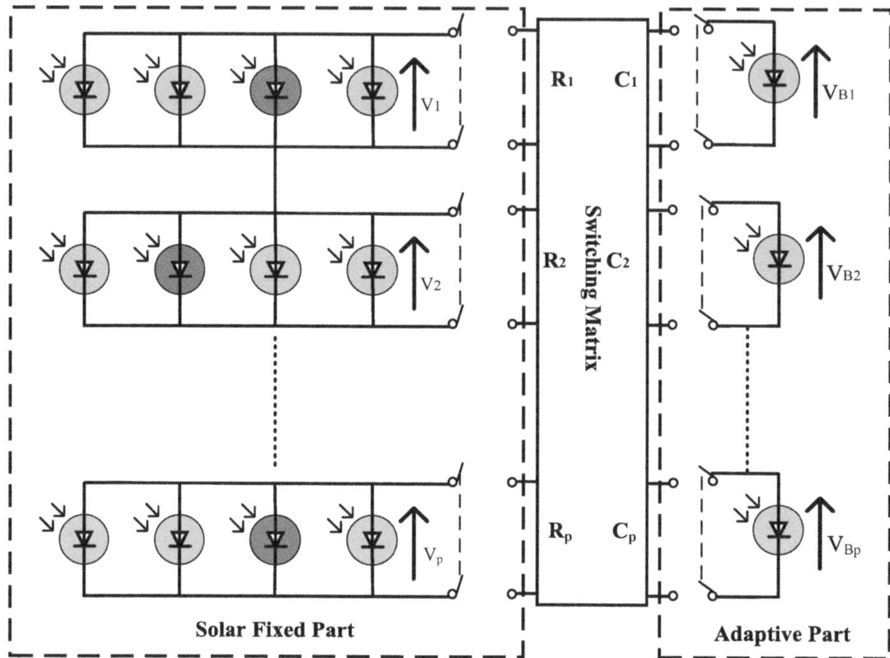

Figure 1.14 Fixed and adaptive part approach for solar cells, the cells with standard irradiation levels are represented in yellow color and less irradiation levels are represented with gray color.

to reconfigure these rows to get into tolerable values. The reconfiguration of these rows is shown in Figs. 1.15(b)–(d). The final reconfiguration matrix is shown in Fig. 1.15(e). In this reconfiguration, all the irradiations lie within the tolerable limits. The final reconfiguration matrix is shown in Fig. 1.15(e). In this matrix, irradiation in all the rows lie within the tolerable limits. Finally, the algorithm looks for optimal reconfiguration based on minimal EI. The biggest problem with this reconfiguration is, out of 16 panels only 4 panels are changed in the final reconfigured array and all the connections for remaining panels are same and when the shade falls on these panels, the output will be same as in that of non-reconfigured array or even lower.

An optimization technique is proposed in [43–46] in order to disperse the concentrated shading on SPV array. The location of panels in an SPV array is unaltered in this approach but the electrical connections are changed. For determining the optimum configuration, the genetic algorithm (GA) is followed. Under shading conditions, this algorithm creates diverse switching arrangements and equalizes the row currents of all rows. [47–51] gives the detailed knowledge about Particle Swarm Optimizations and [52] presents the Electrical Array Reconfiguration through PSO technique to disperse the shading with an optimal reconfiguration. The benefit of this technique upon GA is that in order to attain effective shading dispersion, it needs only

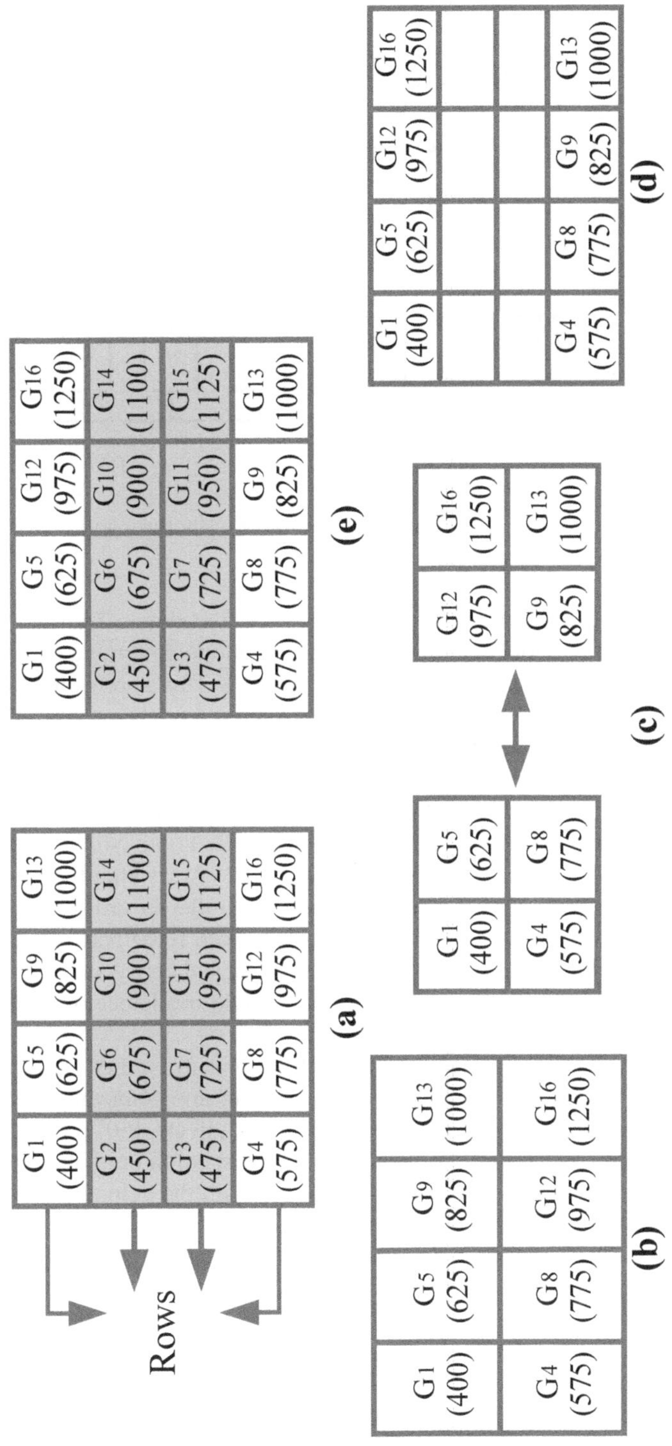

Figure 1.15 (a) Initial Matrix with different irradiations (b)–(d) Reconfiguration using this algorithm (e) Final reconfigurable matrix.

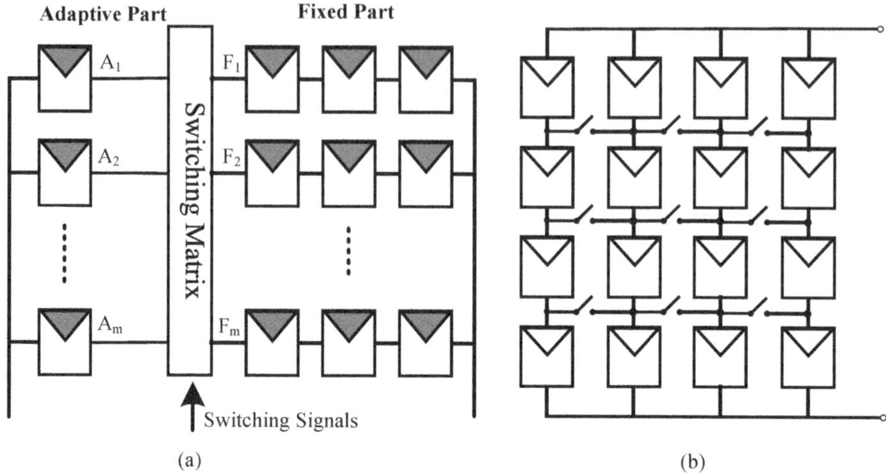

Figure 1.16 (a) Reconfiguration through adaptive structure with fixed part (b) Adaptive PV array reconfiguration.

one time switching. [53,54] presents a variation of this technique. The authors in this study have proposed a unique reconfiguration technique to determine optimum configuration based on a Power Comparison Index (PCI). This approach analyses the maximum powers of all various peaks among Voltage-Power characteristic curves of an SPV array and assures the absolute MPP. Authors in [55] have proposed a Multi Objective Grey Wolf Optimizer proposed for maximizing the generated power. This algorithm is employed for providing optimal switching matrix for a 9×9 SPV Array.

Water pumping system based on new Reconfigurable Photo-Voltaic array (RPV) is proposed in [56]. In this method, the panels in an SPV Array alter their connection layouts from series to parallel to parallel to series parallel in order to operate the motor at different conditions. An adaptive structure with fixed part is cultivated in [57,58]. Figure 1.16(a) depicts the switching matrix tucked between the variable and fixed part. In this method, the measurement of standard irradiance of each string is used to link variable part into fixed part.

A Lambert PV model is established in [59] in order to predict real-time power generation of PV under various conditions. By altering interconnections among PV modules, the proposed model decreases the mismatch effects. [60] presents a variation of this technique. Next, in [61,62] and [63], a robust configuration method based on particle swarm optimization (PSO) is presented. With the help of this algorithm, the optimum configuration through the comparison of maximum power at standard conditions and shaded conditions is obtained. This algorithm promptly undergoes adaptive configuration and identifies an optimum one when any shading or malfunction occurs in PV array. Figure 1.16(b) depicts this adaptive configuration structure. The next version of PSO is explained in [52].

DPVAR strategies demand high end controller, sensor reconfigurations, multiple switching algorithms and complex computations that increase the cost and

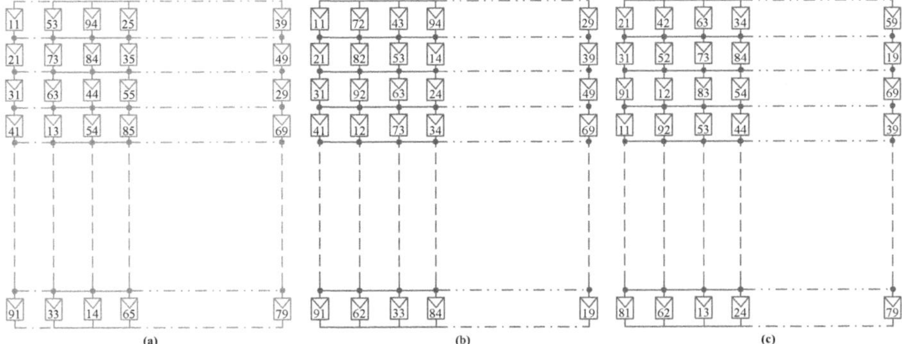

Figure 1.17   (a) Su-Do-Ku reconfiguration (b) Optimal Su-Do-Ku reconfiguration (c) Improved Su-Do-Ku reconfiguration.

complexity of the system. Moreover, the switches in DPVAR strategies even lead to the harmonics because of continuous switching. Abiding by all these disadvantages, the DPVAR strategies are not a feasible solution to use in practice for real-time applications.

1.4.4   STATIC PHOTO-VOLTAIC ARRAY RECONFIGURATION (SPVAR) STRATEGIES

In SPVAR, the physical locations of shaded and unshaded panels are changed, without altering their electrical connections to enhance power output by shade dispersion. Figure 1.2 illustrates the different stages of SPVAR approach. These are One-Time reconfiguration techniques and do not require any sensors, switches and control algorithms as in case of dynamic reconfiguration techniques. The physical locations of the shaded and unshaded panels are changed without altering its electrical connections.

Determining the efficient pattern to disperse the concentrated shade on to the entire array is the most stimulating point. Su-Do-Ku puzzle pattern is employed for the positioning of 9×9 TCT PV array in [64] as depicted in Fig. 1.17(a). Su-Do-Ku is a number placement puzzle which is built on logic. In Su-Do-Ku pattern, the first and second digits in box denote the logic and column numbers, respectively. Distribution of shading effects by altering the physical position of PV panels with same electrical condition is the main purpose of this arrangement. For instance, panel number 53 is in fifth row third column which is relocated to first row third column with this arrangement.

The limitations of this technique are (1) there is a significant increase in wiring when compared with the conventional configuration and (2) under the sub-array, the shading distribution is ineffective. To overcome these limitations, instead of selecting the pattern of user choice, an algorithm for formulation of an optimal Sudoku pattern is proposed [65] under different shading scenarios for a 9×9 SPV Array. Figure 1.17(b) depicts this puzzle arrangement and is demonstrated as follows: In first step, first column is filled with numbers 1 to 9. It is required to have unique numbering in each sub-array so as to fill the next column. In second step, filling of the second column is done by shifting the previous column by three (sub array) and the same procedure is repeated for the subsequent columns also. In third step, fourth column is filled. Filling of the fourth column is done by offset shifting method of the first column. Therefore, the second element begins with one and then goes on. The following two columns are filled employing a shift of three in regard with the previous column, the offset is augmented and filled in the seventh column.

Required shifts are filled in the two subsequent columns which could be demonstrated as

$$n^{th} column = \left\lceil \frac{n-1}{f} \right\rceil + f \times ((n-1) \bmod f) \tag{1.9}$$

where $\lceil \rceil$ is the greatest integer function.

A power increment is observed in Optimal Sudoku pattern to that of Sudoku configuration for each partial shading conditions. However, some disadvantages still follow such as (i) The first column is unaltered and (ii) These patterns have repeated row numbers in diagonal. This also create less output power. The above said disadvantages are fulfilled with Improved Su-Do-Ku Algorithm presented in [66] as shown in Fig. 1.17(c).

Whenever there is space constraint and angle of sun elevation is low, then there is mutual shading among the PV panels. So, in order to tackle the problem of mutual shading another method was introduced in [67, 68]. In the real-time PV plant, the MSH generally appears due to the motionless solar collectors and usually in locations of limited solar land and low sun elevation angle as shown in Fig. 1.18. The mutual shading pattern can be estimated by considering position of sun, geographical position of power plant and installation conditions of the PV panels. This method disperses the mutual shading better compared to normal sudoku arrangement.

In [69] for the effect of Moving Shadow (passing cloud) conditions, Su-Do-Ku riddle puzzle pattern is demonstrated. Various PV array configurations use this pattern arrangement such as Series-Parallel (SP), Bridge-Link (BL), Honeycomb (HC), Total Cross Tied (TCT) and Ladder (LDR). LDR arrangement has the links of the alternate rows via cross ties. Subsequently in [70] to distribute the shading effects, a similar arrangement is demonstrated. An alternative for this arrangement is demonstrated in [71].

In [72], hybrid PV array configurations and non-symmetrical patterns are proposed in order to distribute shading effects. Figure 1.19(a) and (b) show the hybrid configurations like SP-TCT and BL-TCT, respectively. Figs. 1.19(c)–(f) show the proposed pattern as well as arrangements of NS-1 (Non-Symmetrical-1) and NS-2 (Non-Symmetrical-2) reconfigurations, respectively. NS-1 configuration is built by organizing PV panels in the ascending order in the first column and then, from column 2, number 2 is added to the previous column. For the formation of NS-2 pattern, the same method is repeated with the addition of 3 rather than 2. In this approach, the power loss is minimal in the proposed NS-1 and NS-2 configurations when compared to other configurations.

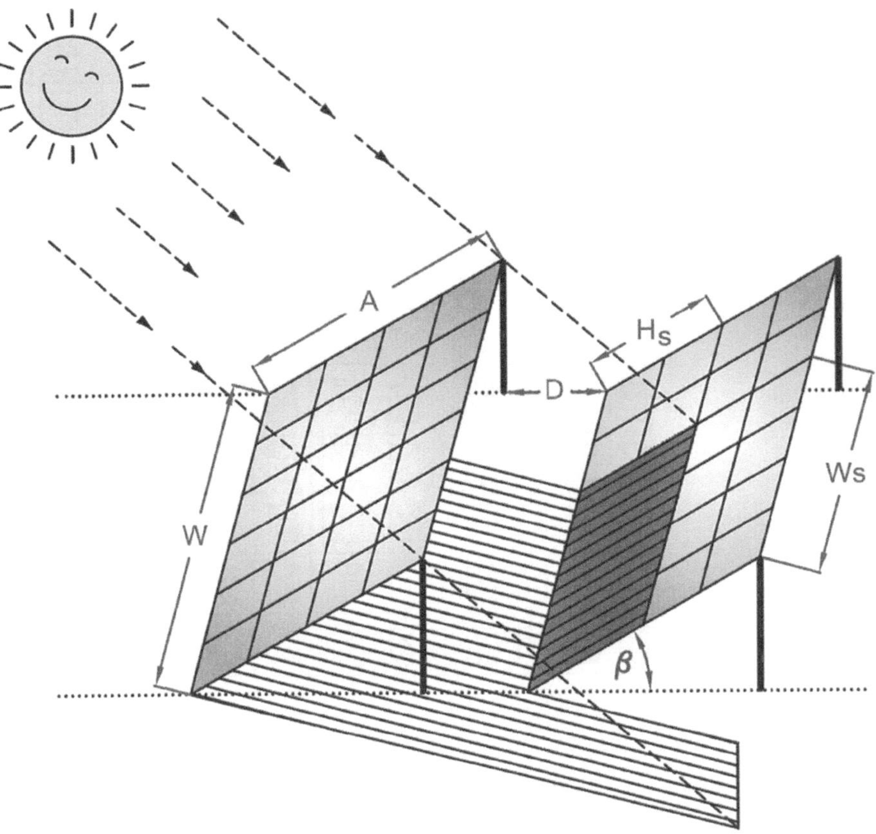

Figure 1.18 Occurrence of mutual shading between collectors [67, 68].

With Sudoku technique there was a disadvantage that it can be applied to only 9×9 PV array so to overcome this Magic Square (MS) technique was introduced which can be implemented for any $N \times N$ PV array. The Magic Square technique follows a condition under which the sum of the currents should be same in rows, columns and diagonal as far as possible. This means that the reconfiguration takes place in such a way that irradiation level in rows, columns and diagonals are maintained almost same. A Magic-Square (MS) puzzle pattern is taken up for various hybrid PV array configurations in [73]. MS refers to square grid arrangement inside the array. Each square grid has exclusive pattern of numbering in them. The addition of numbers in each row, column and diagonal provides the same number. The MS formation of hybrid configurations like rearranged-total cross tied (RTCT), rearranged series parallel-total cross tied (RSP-TCT), rearranged bridge link-total cross tied (RBL-TCT) and rearranged bridge link-honey comb (RBL-HC) are depicted in Fig. 1.20(a)–(d), respectively. The results show that the MS configuration holds minimum power loss and higher fill factor.

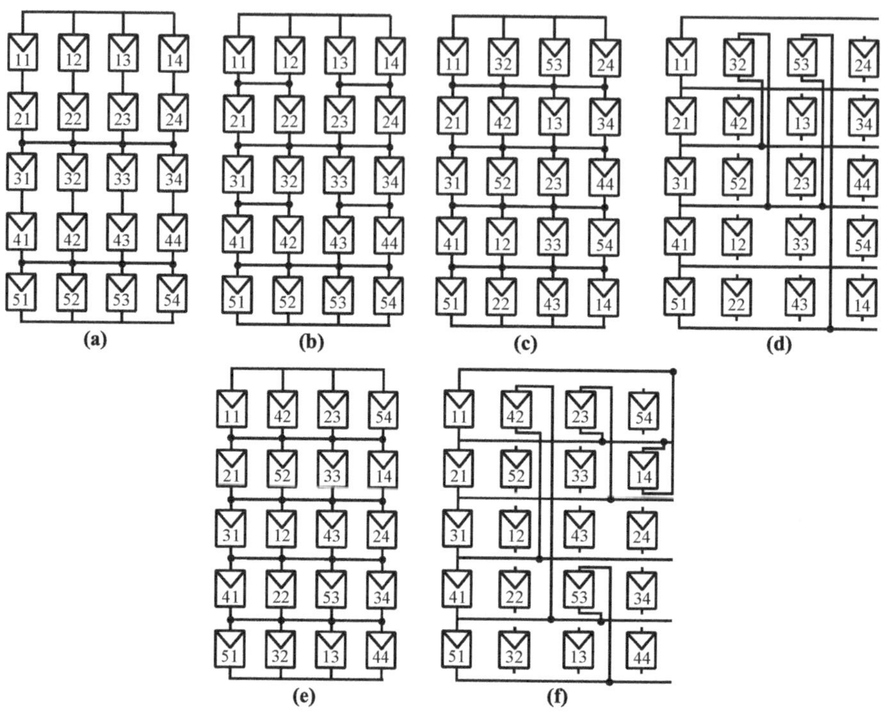

Figure 1.19 Different interconnection techniques (a) SP-TCT (b) BL-TCT (c) Proposed NS-1 pattern (d) Proposed NS-1 arrangement (e) Proposed NS-2 pattern (d) Proposed NS-2 arrangement.

The reconfiguration for a 3×3 PV array using MS technique is represented in [74–78] represented Magic Square against basic interconnection schemes. The result of the comparison confirmed the MS technique is better than basic interconnection schemes in terms of improving the output power under different shading conditions. The work in [79] represented MS technique and this technique helped in better dispersion of shading condition improving the power and reducing the mismatch losses. It also aids in MPPT as it reduces the mismatch current which helps to smoothen the characteristics (P-V) curves. The drawback of this technique is that it cannot be applied to large PV plants. To overcome this drawback new improved version of MS technique was presented in [80, 81] called Magic Square-Enhanced Technique (MS-ET). Under this technique the PV array is considered as collection of groups and each group is under same level of solar irradiation. So, this helps to increase output power and reduce the mismatch problem.

Futoshiki-based puzzle pattern is developed by the authors in [87] for improving power generation in the effect of partial shading. Futoshiki refers to a logic-based puzzle with inequality between two adjacent numbers. The numbers 1 to 5 are arranged in this pattern so that no number is repeated in each row and column

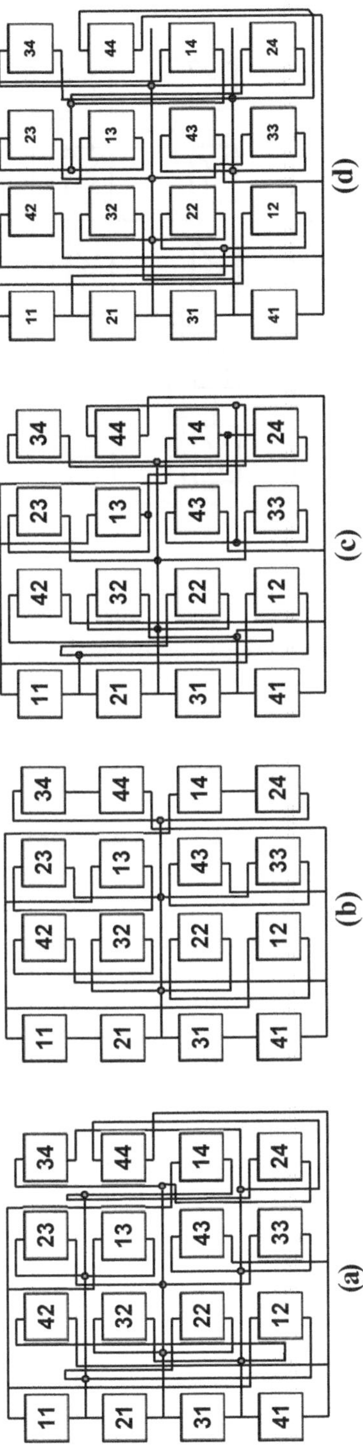

Figure 1.20 Physical relocation with Magic Square reconfiguration (a) RTCT (b) RSP-TCT (c) RBL-HC (d) RBL-HC [29].

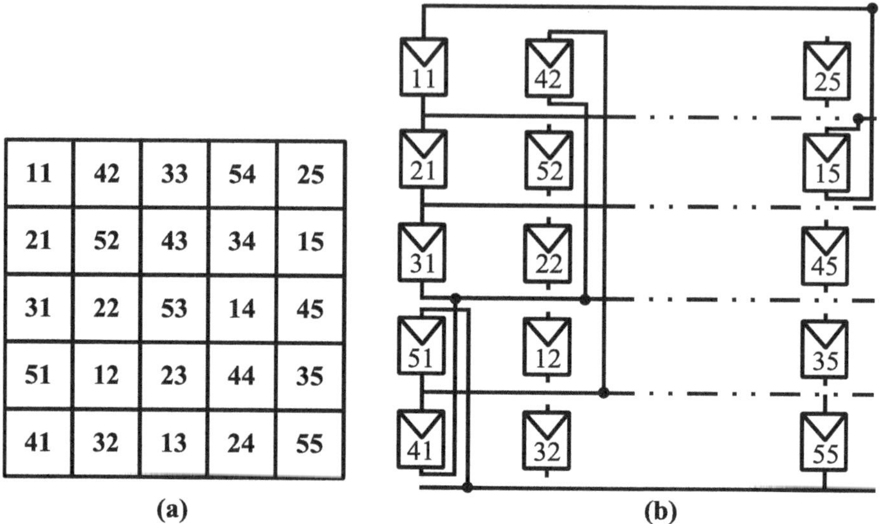

Figure 1.21   (a) Futoshiki pattern (b) Arrangement.

of a square grid (Fig. 1.21). While arranging the numbers in this grid the initially specified inequality constraint has to be fulfilled between two adjacent numbers. Regardless of matter how efficient it is, the implementation of this function cannot be reduced or extended beyond a 5×5 array. Furthermore, this approach includes several solutions, and determining the best pattern for various shading circumstances is a time-consuming and hard operation.

Reconfiguration based on Dominance Square (Fig. 1.22) and Competence Square (Fig. 1.23) was proposed in [88, 89], respectively, to attain maximum power during partial shading. However, these kinds of rearrangements are limited to tiny SPV panels and result in poor output power under certain frequent column shadowing situations. The major disadvantages with these methods are that they are applicable to 9×9 and 5×5 respective sizes of arrays only. Another reconfiguration was proposed in [90] to mitigate the shading effects for a 9×9 array with two-phase array reconfiguration. In the first phase the sub arrays in an array is relocated and in the second phase the panels in the sub arrays are relocated to achieve final shade dispersion pattern. This method is compared between TCT and Su Do Ku and two-phase reconfiguration gives maximum enhanced parameters than the remaining two. But the problem of shade dispersion is a well made issue yet.

A technique was presented in [98] called as SD-PAR based on relocation of shaded and unshaded panels. The relocation with this SD-PAR is represented in Fig. 1.24. This method is implemented on a 3×3 SPV Array and compared with SP, TCT and BL configurations and found that SD-PAR exhibits high performance parameters than the other configurations. A one-time reconfiguration based on skyscraper puzzle is proposed in [95] to disperse the shade thereby enhancing the output power. Skyscraper puzzle is based on arranging the buildings according to its

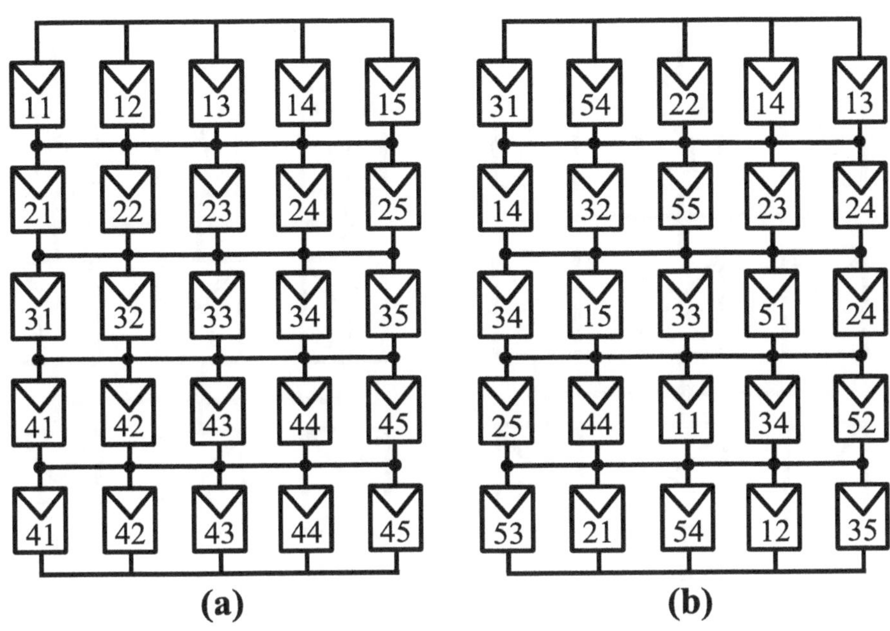

Figure 1.22 (a) TCT pattern (b) Dominance Square reconfiguration pattern.

11	12	13	14	15	16	17	18	19
21	22	23	24	25	26	27	28	29
31	32	33	34	35	36	37	38	39
41	42	43	44	45	46	47	48	49
51	52	53	54	55	56	57	58	59
61	62	63	64	65	66	67	68	69
71	72	73	74	75	76	77	78	79
81	82	83	84	85	86	87	88	89
91	92	93	94	95	96	97	98	99

(a)

56	66	76	86	96	16	26	36	46
37	47	57	67	77	87	97	17	27
18	28	38	48	58	68	78	88	98
89	99	19	29	39	49	59	69	79
61	71	81	91	11	21	31	41	51
42	52	62	72	82	92	12	22	32
23	33	43	53	63	73	83	93	13
94	14	24	34	44	54	64	74	84
75	85	95	15	25	35	45	55	65

(b)

Figure 1.23 (a) TCT pattern (b) Competence Square reconfiguration pattern.

heights with the key. The reconfiguration with this Skyscraper pattern is shown in Fig. 1.25. These results show the supremacy of skyscraper puzzle pattern when compared with TCT, Su Do Ku and DS configurations. The downsides of these methods are its additional power loss caused by the cables (or wires) and these can only be implemented to respective size of array only. Table 1.2 presents the summary of SP-VAR with different performance parameters.

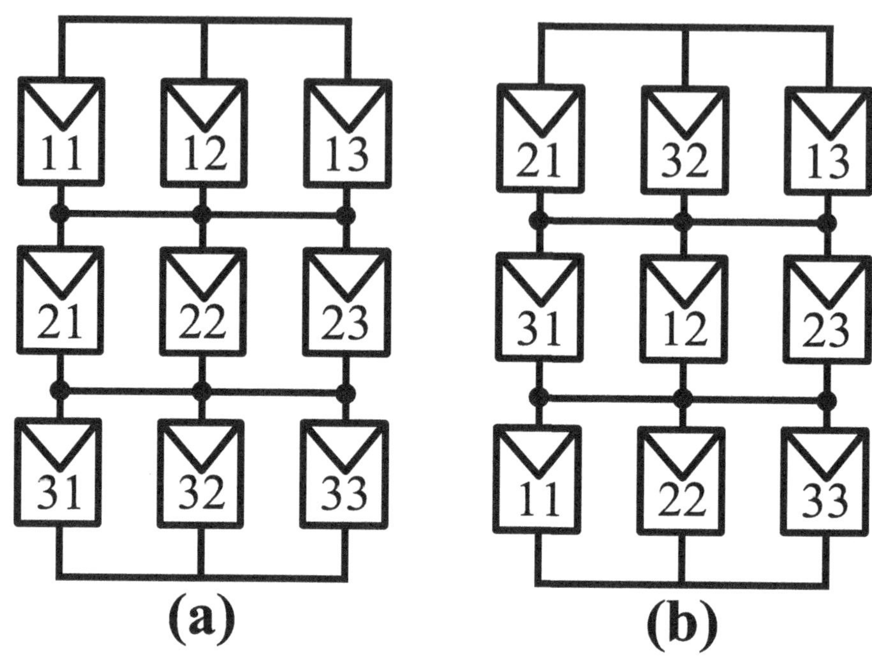

Figure 1.24   (a) TCT pattern (b) SD-PAR reconfiguration.

11	12	13	14	15	16	17	18	19
21	22	23	24	25	26	27	28	29
31	32	33	34	35	36	37	38	39
41	42	43	44	45	46	47	48	49
51	52	53	54	55	56	57	58	59
61	62	63	64	65	66	67	68	69
71	72	73	74	75	76	77	78	79
81	82	83	84	85	86	87	88	89
91	92	93	94	95	96	97	98	99

(a)

41	92	63	24	35	86	17	78	59
21	72	53	44	15	96	87	68	39
81	32	13	64	75	56	47	28	99
61	42	83	34	95	16	27	58	79
11	22	73	54	65	46	97	38	89
31	52	93	74	85	26	67	48	19
71	12	43	84	55	66	37	98	29
51	82	23	94	45	36	77	18	69
91	62	33	14	25	76	57	88	49

(b)

Figure 1.25   (a) TCT pattern (b) Skyscraper reconfiguration.

The output power in SPVAR is a bit low when compared with DPVAR strategies but a huge savings in terms of monitory and maintenance is observed to compensate the power. Because of the above said advantages, the SPVAR strategies are used in practice.

Table 1.2
Summary of SPVAR Strategies

Reference	Reconfiguration	Array Size	Shading Used	Complexity
[64]	Su Do Ku	9×9	SW, LW, SN, LN	High
[65]	Optimal Su Do Ku	9×9	SW, LW, SN, LN	High
[66]	Improved Su Do Ku	9×9	4 groups of shadings on each corner of array	High
[69]	Riddle Su Do Ku	6×6 in sub grids of 3×2	Different Shadings	High
[82]	SDS	$3 \times 3, 7 \times 7$	Different Shadings	High
[83]	SDP Scheme	$3 \times 3, 5 \times 5$	SW, LW, SN, LN	Medium
[70]	Zig-Zag	4×3	Single row, Double row, Corner, Diagonal	Medium
[84]	Adjacent Shifting Pattern	9×9	SW, LW, SN, LN	High
[85]	Latin Square Puzzle	9×9	SW, LW, SN, LN	High
[22]	Odd-Even Structure	4×4	Arbitrary	Medium
[86]	Odd-Even-Prime Structure	9×9	4 groups of shadings on each corner of array	Medium
[87]	Futoshiki	5×5	SW, LW, SN, LN	Medium
[88]	Dominance Square	5×5	SW, LW, SN, LN	High
[89]	Competence Square	9×9	SW, LW, SN, LN	High
[80]	Cross Diagonal View	9×9	Inner Block, Diagonal, Corner	Medium

(Continued on next page)

[90]	Two-Phase array reconfiguration	9×9	SW, LW, SN, LN	High
[72]	Non-Symmetrical-1, 2	5×4	Incremental	Medium
[91]	Latin Square Puzzle	4×4	Incremental	Medium
[73]	Magic Square	4×4	Incremental	Low
[92]	Non-Symmetrical	6×4	Incremental	Low
[76]	Magic Square	$3 \times 3, 6 \times 6$	SW, LW, SN, LN	Medium
[93]	Fixed Electrical Reconfiguration	$3 \times 4, 6 \times 6$	SW, LW, SN, LN, Diagonal	Low
[94]	Column Indexed	9×9	SW, LW, SN, LN	Low
[95]	Skyscraper	9×9	SW, LW, SN, LN	High
[74]	Magic Square	4×4	SW, LW, SN, LN	Medium
[96]	Improvised Magic Technique	$3 \times 3, 9 \times 9$	SW, LW, SN, LN	Medium
[97]	Magic Sudoku	9×9	SW, LW, SN, LN	High
[98]	SD-PAR	3×3	SW, LW, SN, LN	Medium
[99]	Skyscraper	6×6	Row, Column and Center	High

2 Array Reconfiguration with Tom-Tom Puzzle Pattern

2.1 INTRODUCTION

Partly shading on panels is an unavoidable challenge, and management of partial shaded panels has arisen as a key issue. Because the electrical properties of shaded and unshaded panels change, the energy harvest is reduced. To maximize energy harvest under partial shadowing circumstances, an unique reconfiguration based on a Mathematical Tom-Tom puzzle pattern is developed and executed on a 5×5 array. The primary rationale for using this technology is its effectiveness in dispersing shadow and improving array utilization in terms of power by lowering mismatch losses under partial shading circumstances. The suggested shadow dispersion using Tom-Tom reconfiguration pattern has the benefit of being applicable to all types of shading circumstances.

The rest of the chapter is organized as follows: Section 2.2 describes the approach for array reconfiguration. Section 2.3 discusses about the results when compared with the conventional configurations. Finally, the chapter is summarized in Section 2.4

2.2 TOM-TOM RECONFIGURATION TOPOLOGY

To increase the output power of the SPV array, a hand-made mathdoku puzzle pattern known as the Tom-Tom puzzle pattern is utilized. This pattern is used to reconfigure the shaded and unshaded panels of the array in order to minimize the mismatch of row currents. This is accomplished by dispersing the concentrated shade across the entire array for each and every shading condition. The renumbering of the panels in the SPV array is the first step in this pattern. The reorganization is carried out in such a manner that different nodes get an equal quantity of the current at their respective entry points. The photovoltaic panels, which are normally arranged in a single row in traditional configurations, will now be arranged in many rows under the suggested configuration. This helps to mitigate the effect of shading by ensuring that it is evenly applied over the whole array. Because of this, the current that is flowing from a particular node may be improved by using the Tom-Tom configuration that has been provided in order to increase the output power production under situations of partial shade.

A Tom-Tom puzzle's primary rule is that the $N \times N$ grid should be filled with numbers from 1 to N such that no number is repeated in any row or column. The sub-grids are also included in the N to N grid. On the left-top corner of each sub-grid is a number and the operation to be done on the numbers of the relevant grid.

A sub-numbers grid's should be filled in such a way that the specified number may be obtained by utilizing these numbers and the mathematical operation specified

DOI: 10.1201/9781003285830-2

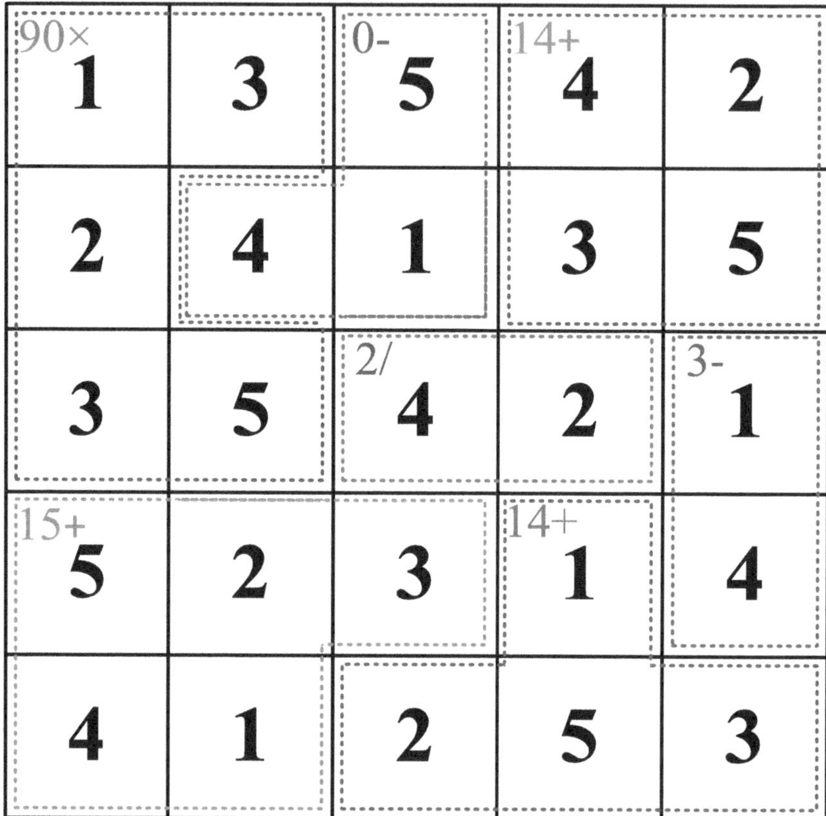

Figure 2.1 Tom-Tom puzzle pattern.

on them. In Fig. 2.1, for example, a patch of Tom-Tom puzzle design with '30×' is depicted. '30×' specifies that when each integer in that specific cell is multiplied in that specific location, the resulting should be 30, i.e., $1 \times 3 \times 2 \times 5 = 30$. The numbers may be repeated inside the sub grid as long as the primary rule is not broken. Number '3', for example, is repeated in the area where '90×' is acquired. When doing subtraction and division, begin with the greatest number of the sub-grid.

A 5×5 Tom-Tom puzzle design is developed in this study to spread the shadow under partial shading situations. This puzzle design is created by combining a backtracking algorithm with the mathematical procedure indicated in the sub-grid. Backtracking is an important method for solving constraint issues. The main limitation in this puzzle design is not to duplicate the numerals 1 to 5 in rows and columns. The sub-grids may have numerical repetition. The procedures for finishing the problem and acquiring the redesigned puzzle pattern for Tom-Tom configuration are as

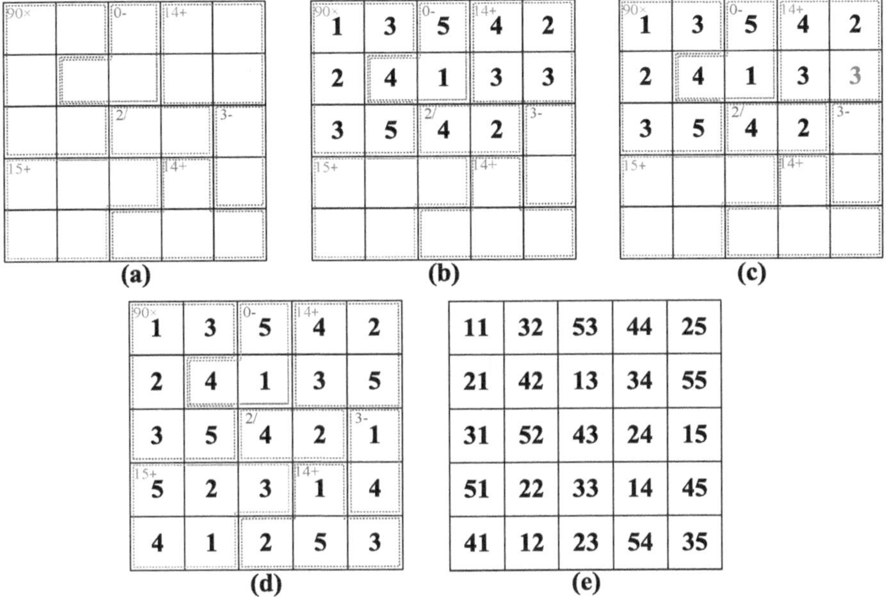

Figure 2.2 (a) Array taken for consideration with mathematical operations on the left top corner of every sub-grids (b) Filling up the empty cells with numbers from 1 to 5 (indicated in black color) (c) Failing of condition for the Tom-Tom puzzle pattern (indicated in red color) (d) Fully completed puzzle pattern by using back tracking (e) Final Tom-Tom reconfiguration used for the study.

follows:

- Choose the Tom-Tom Puzzle Pattern and fill the remaining cells with integers ranging from 1 to 5 (Fig. 2.2(a))
- If the requirements of no duplication of numbers in rows and columns and mathematical operation of numbers in sub-grid are met, then fill all the vacant cells in the same manner to create the unique answer (Fig. 2.2(b)).
- Otherwise, use the backtracking approach to alter the placement of the repeated ones and finish the recursive testing to generate a final unique puzzle design (Fig. 2.2(c) and Fig. 2.2(d)).
- Adding the column number to the final obtained Tom-Tom Puzzle pattern to build a Tom-Tom reconfigurable array that is utilized for SPV array (Fig. 2.2(e)).

Figure 2.3 depicts the structural arrangement of Fig. 2.2(e). The panels that were in one row of the TCT arrangement have been moved to another. In the TCT layout, for example, panel number '43' (fourth row, third column) is physically and electrically linked between '33' (third row third column) and '53' (fifth row, third column). This '43' panel is relocated to the third row, third column in the Tom-Tom layout

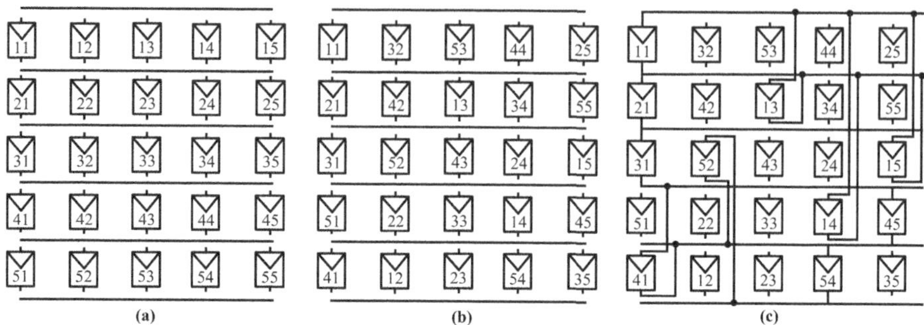

Figure 2.3 Structural arrangement of Tom-Tom reconfiguration (a) Initial panel organization; (b) Reconfigured panel organization (c) Final connections with Tom-Tom configuration.

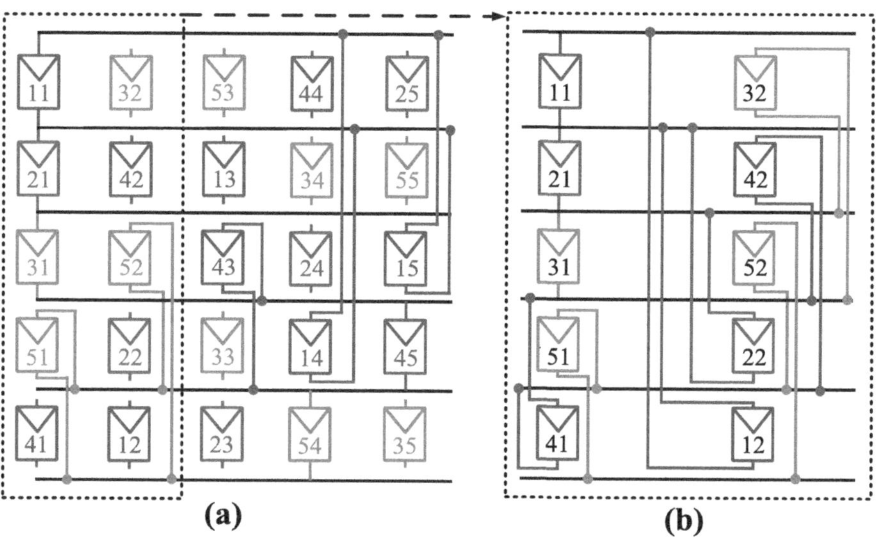

Figure 2.4 (a) Panel arrangement in suggested Tom-Tom setup (b) Clearly show the first two columns of panel reconfiguration with the same electrical connections.

(Fig. 2.3(b)), but the electrical connections are the same as those illustrated in Fig. 2.3(c) for TCT. Figure 2.4 depicts a zoomed-in view of the first and second columns for the modified array.

2.3 RESULTS AND DISCUSSIONS

The Tom-Tom and conventional configurations are compared with different non-uniform and incremental moving shading conditions. The location of global

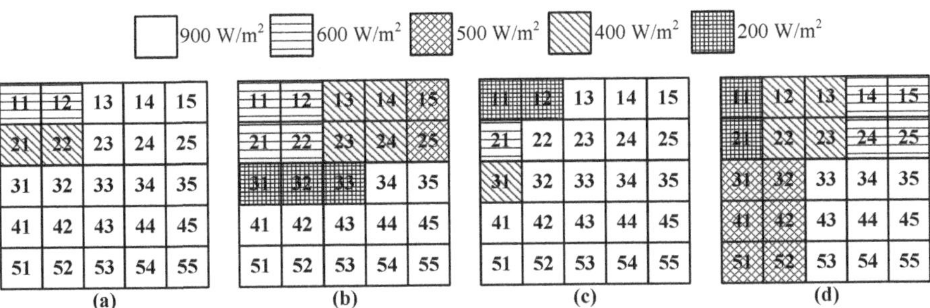

Figure 2.5 Shading profile for 5×5 under Non-Uniform shading conditions used in this work.

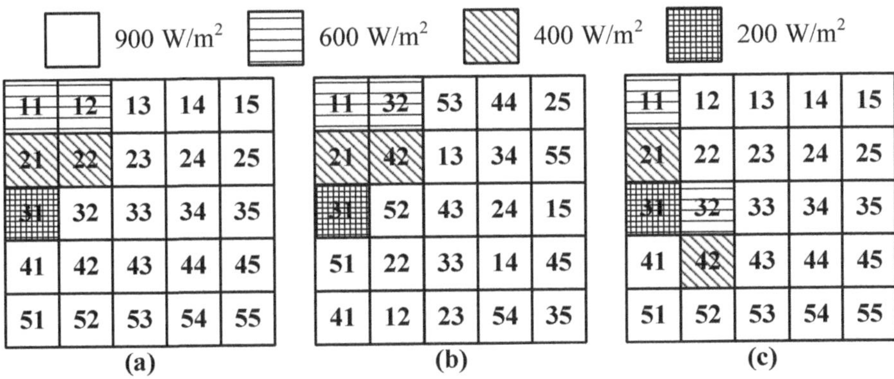

Figure 2.6 Pattern for Short Narrow static shading conditions (a) TCT configuration (b) Tom-Tom configuration (c) Dispersing the shade with Tom-Tom configuration.

maximum power point in Voltage-Power characteristics of SPV array is discussed along with absolute mismatch power loss for each case.

2.3.1 NON-UNIFORM SHADING CONDITIONS

In this work, four static shading conditions are used in this work as given in Fig. 2.5.

2.3.1.1 Short Narrow Shading Conditions

In this scenario, just a small number of panels are shaded. In this scenario, four unique groups of solar irradiation are studied. As indicated in Fig. 2.6, group one gets 200W/m^2 (most dark shade), group two receives 400W/m^2 (reduced intensity of light), group three receives 600W/m^2 (less shaded), and group four receives 900W/m^2 (highest illumination).

It is necessary to analyze current generated in each row of the array while calculating the theoretical location of the GMPP in TCT setup. The current produced at an irradiation G is given by $I = xI_m = (G/G_o)I_m$.

The following is how the current looks in row 1

$$I_{R1} = x_{11}I_{11} + x_{12}I_{12} + x_{13}I_{13} + x_{14}I_{14} + x_{15}I_{15} \tag{2.1}$$

where, $x_{11} = (G_{11}/G_0)$; G_{11} is the insolation received on panel 11. Substituting all the values, the row 1 current is:

$$I_{R1} = 2 \times 0.6I_m + 3 \times 0.9I_m = 3.9I_m \tag{2.2}$$

In row 2, two of the panels are exposed to an illumination of $400\text{W}/\text{m}^2$, while the other three are hit with $900\text{W}/\text{m}^2$. The current that is flowing across is provided by

$$I_{R2} - 2 \times 0.4I_m \mid 3 \times 0.9I_m = 3.5I_m \tag{2.3}$$

Only one of the panels in row 3 is exposed to the $300\text{W}/\text{m}^2$ whereas the other four are subjected to the $900\text{W}/\text{m}^2$ value. The current that is flowing across is provided by

$$I_{R3} = 1 \times 0.3I_m + 4 \times 0.9I_m = 3.9I_m \tag{2.4}$$

Rows 4 and 5 are completely lighted. The currents are provided by

$$I_{R4} = I_{R5} = 5 \times 0.9I_m = 4.5I_m \tag{2.5}$$

For Tom-Tom configuration shown in 2.6(c), because of the shade dispersion, the four shaded panels are shaded in four different rows. So, the currents are given by:

For row 1, only one panel is shaded with $600\text{W}/\text{m}^2$ and remaining all panels receive an irradiation level of $900\text{W}/\text{m}^2$.

$$I_{R1} = 1 \times 0.6I_m + 4 \times 0.9I_m = 4.2I_m \tag{2.6}$$

Rows 2 and 4 receive same irradiation levels, for configuration, one panel is subjected to $400\text{W}/\text{m}^2$ and remaining panels are subjected to $900\text{W}/\text{m}^2$. So, the current in Row 2 is same as that of Row 4 which is given as

$$I_{R2} = I_{R4} = 1 \times 0.4I_m + 4 \times 0.9I_m = 4I_m \tag{2.7}$$

For Row 3, only one panel is shaded with $600\text{W}/\text{m}^2$ and another with $200\text{W}/\text{m}^2$, remaining all panels receive an irradiation level of $900\text{W}/\text{m}^2$.

$$I_{R3} = 1 \times 0.3I_m + 1 \times 0.6I_m + 3 \times 0.9I_m = 3.6I_m \tag{2.8}$$

In Row 5 all the panels are unshaded and are subjected to $900\text{W}/\text{m}^2$

$$I_{R5} = 5 \times 0.9I_m = 4.5I_m \tag{2.9}$$

Table 2.1

Maximum Power Point Position in Total Cross Tied and Tom-Tom Configurations for Short Narrow Shade Conditions

Total Cross Tied Configuration			Tom-Tom Configuration		
Current(A)	Voltage(V)	Power(W)	Current(A)	Voltage(V)	Power(W)
I_{R2} $3.5I_m$	$5V_m$	$17.5V_mI_m$	I_{R3} $3.6I_m$	$5V_m$	$18V_mI_m$
I_{R1} $3.9I_m$	$4V_m$	$15.6V_mI_m$	I_{R2} $4I_m$	$4V_m$	$16V_mI_m$
I_{R3} $3.9I_m$	$3V_m$	$11.7V_mI_m$	I_{R4} $4I_m$	$3V_m$	$12V_mI_m$
I_{R4} $4.5I_m$	$2V_m$	$9V_mI_m$	I_{R1} $4.2I_m$	$2V_m$	$8.4V_mI_m$
I_{R5} $4.5I_m$	V_m	$4.5V_mI_m$	I_{R5} $4.5I_m$	V_m	$4.5V_mI_m$

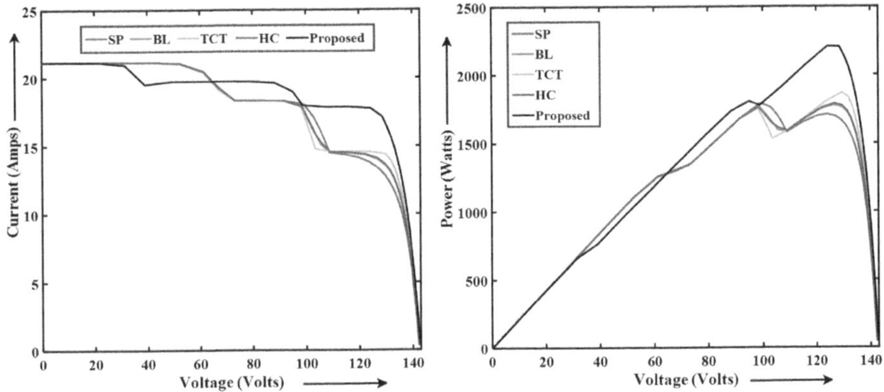

Figure 2.7 Voltage-Current and Voltage-Power characteristics for SP, BL, TCT, HC, and Tom-Tom configurations for Short Narrow shading conditions.

The theoretical values for Currents (in the order in which the panels are bypassed), Voltages and Powers to find the location of GMPP for conventional and configuration for each and every row under short narrow shading conditions are tabulated in Table 2.1. The Voltage-Current and Voltage-Power characteristics for all conventional and Tom-Tom are shown in Fig. 2.7. It can be observed from Fig. 2.7, that under short narrow shading conditions, the conventional configurations possess the multiple maximum power points. By dispersing the shading pattern with the Tom-Tom configuration throughout the array, the multiple peaks were eradicated thereby only single maximum power point is observed in the characteristics curve. The power improvement is obtained due to re-shuffling of the shaded PV panels throughout the array which tends to reduce the mismatch losses.

The results accounting the maximum generated output power (W), power loss (W) and power loss (%), for SP, BL, TCT, HC, and Tom-Tom topologies under Short Narrow shading conditions are given in Table 2.2. Under no shading i.e., at STC 5×5

Table 2.2

Generated Output Power (W) and Overall Power Loss Caused by Shading for Different Configurations Under Short Narrow Shading Condition

Topology	Power (W)	Power Loss (W)	Power Loss (%)
SP	1790	960	34.90
BL	1775	975	35.45
TCT	1866	884	32.14
HC	1786	964	35.05
TT	**2206**	**544**	**19.78**

array yields an output power of 2750W. SP yields an output power of 1790W with an overall shading loss of 960W (34.90%). The mostly used benchmark configuration TCT yields the power of 1866W with a shading loss of 884W (32.14%). BL and HC configurations yield an output power of 1775W and 1786W respectively, with a power loss of 975W (35.45%) and 964W (35.05%), whereas in the Tom-Tom configuration, the power yield has increased to 2206W thereby, minimizing the power loss to 544W (19.78%). Hence, it is validated that the configuration is superior under Short Narrow shading condition compared with SP, BL, TCT, and HC.

2.3.1.2 Short Wide Shading Conditions

Under this shading condition, five different types of irradiance levels are applied to the Solar PV Array, which are $900W/m^2$, $600W/m^2$, $500W/m^2$, $400W/m^2$, and $200W/m^2$. The pattern in which shading occurs conventionally has been shown in Fig. 2.8.

The current calculations for each row under the TCT configuration are given as,

$$
\begin{aligned}
I_{R1} &= 1 \times 0.5I_m + 2 \times 0.4I_m + 2 \times 0.6I_m = 2.5I_m \\
I_{R2} &= I_{R1} = 2.5I_m \\
I_{R3} &= 3 \times 0.2I_m + 2 \times 0.9I_m = 2.4I_m \\
I_{R4} &= I_{R5} = 5 \times 0.5I_m = 4.5I_m
\end{aligned}
\tag{2.10}
$$

The theoretical current output for each row in the shade dispersed Tom-Tom configuration shown in Fig. 2.8(c) is as follows

$$
\begin{aligned}
I_{R1} &= 1 \times 0.4I_m + 1 \times 0.6I_m + 3 \times 0.9I_m = 3.7I_m \\
I_{R2} &= 1 \times 0.5I_m + 3 \times 0.9I_m + 1 \times 0.6I_m = 3.8I_m \\
I_{R3} &= 1 \times 0.5I_m + 1 \times 0.4I_m + 2 \times 0.9I_m + 1 \times 0.6I_m = 3I_m \\
I_{R4} &= I_{R3} = 3I_m \\
I_{R5} &= 1 \times 0.5I_m + 1 \times 0.4I_m + 2 \times 0.9I_m + 1 \times 0.2I_m = 2.9I_m
\end{aligned}
\tag{2.11}
$$

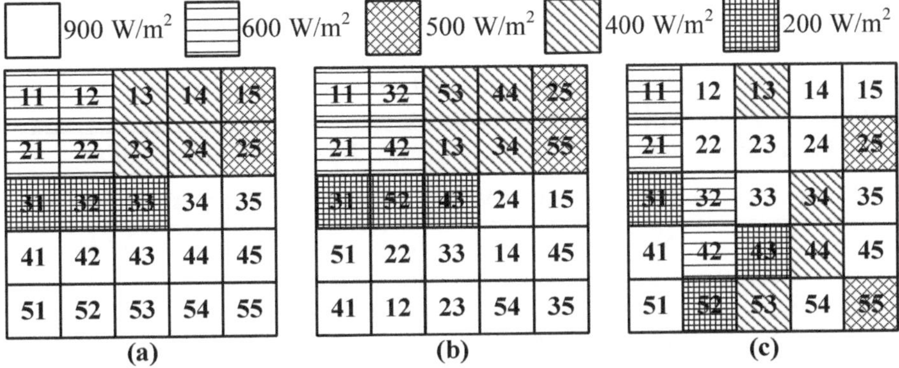

Figure 2.8   Pattern for Short Wide static shading conditions (a) TCT configuration (b) Tom-Tom configuration (c) Dispersing the shade with Tom-Tom configuration.

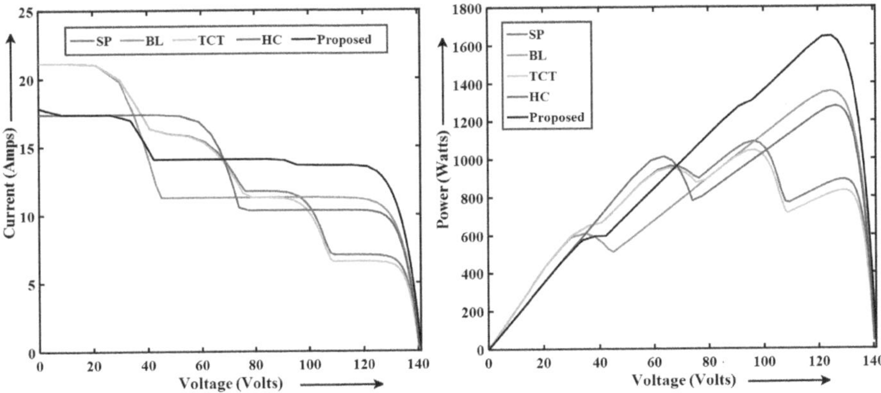

Figure 2.9   Voltage-Current and Voltage-Power characteristics for SP, BL, TCT, HC, and Tom-Tom configurations for Short Wide shading conditions.

The theoretical values for currents (in the sequence under which the panels are by-passed), voltages and powers are employed to locate the GMPP for conventional and configuration under short wide shading conditions is tabulated in Table 2.3. Figure 2.9 shows the Voltage-Current and Voltage-Power characteristics for all conventional and configurations. It is clearly seen that the shading on the panel causes disruption in the Voltage-Current and Voltage-Power characteristic curves which in fact led to obtaining the multiple power points. The configuration with less disturbances yields more output power than the remaining configurations because the concentrated shaded panels are distributed throughout the entire array.

The power got incremented from 1045W to 1645W from SP to Tom-Tom configuration, 1357W to 1645W from TCT to Tom-Tom configuration, 1092W and 1276W from BL, HC configurations to Tom-Tom configuration. The power loss occurred

Table 2.3

Maximum Power Point Position in Total Cross Tied and Tom-Tom Configurations for Short Wide Shade Conditions

Total Cross Tied Configuration			Tom-Tom Configuration		
Current(A)	Voltage(V)	Power(W)	Current(A)	Voltage(V)	Power(W)
I_{R3} 2.4I_m	5V_m	12V_mI_m	I_{R5} 2.9I_m	5V_m	14.5V_mI_m
I_{R1} 2.5I_m	4V_m	10V_mI_m	I_{R3} 3I_m	4V_m	12V_mI_m
I_{R2} 2.5I_m	3V_m	7.5V_mI_m	I_{R4} 3I_m	3V_m	9V_mI_m
I_{R4} 4.5I_m	2V_m	9V_mI_m	I_{R1} 3.7I_m	2V_m	7.4V_mI_m
I_{R5} 4.5I_m	1V_m	4.5V_mI_m	I_{R2} 3.8I_m	1V_m	3.8V_mI_m

Table 2.4

Generated Output Power (W) and Overall Power Loss Caused by Shading for Different Configurations under Short Wide Shading Condition

Topology	Power (W)	Power Loss (W)	Power Loss (%)
SP	1045	1705	62
BL	1092	1658	60.29
TCT	1357	1393	50.65
HC	1276	1474	53.6
TT	**1645**	**1105**	**40.18**

during shading (W & %) for SP, BL, TCT, HC, and configurations are 1705W (62%), 1658W (60.29%), 1393W (50.65%), 1474W (53.6%), and 1105W (40.18%), respectively, which are depicted in Table 2.4.

2.3.1.3 Long Wide Shading Conditions

In this shading condition, maximum number of panels are shaded with different illumination levels as $900W/m^2$, $600W/m^2$, $500W/m^2$, $400W/m^2$ and $200W/m^2$. This static shading condition which covers almost 64% of the array is represented in Fig. 2.10 for conventional configuration. Under this type of shading conditions there is an adverse effect on SPV Array as it limits its ability to produce output power to only 30% of the maximum output possible.

The current calculations for each row under TCT configuration are given as

$$I_{R1} = I_{R2} = 0.2I_m + 2 \times 0.4I_m + 2 \times 0.6I_m = 2.2I_m$$
$$I_{R3} = I_{R4} = I_{R5} = 2 \times 0.5I_m + 3 \times 0.9I_m = 3.7I_m$$

$$(2.12)$$

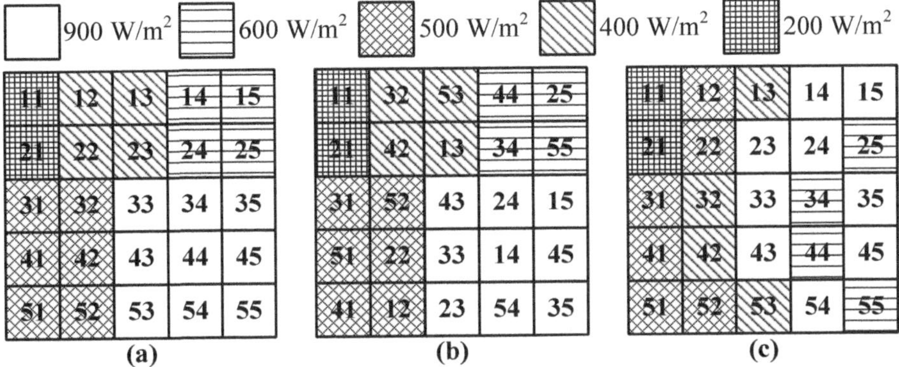

Figure 2.10 Pattern for Long Wide static shading conditions (a) TCT configuration (b) Tom-Tom configuration (c) Dispersing the shade with Tom-Tom configuration.

Table 2.5

Maximum Power Point Position in Total Cross Tied and Tom-Tom Configurations for Long Wide Shade Conditions

Total Cross Tied Configuration			Tom-Tom Configuration		
Current(A)	Voltage(V)	Power(W)	Current(A)	Voltage(V)	Power(W)
I_{R1} $2.2I_m$	$5V_m$	$11V_mI_m$	I_{R1} $2.9I_m$	$5V_m$	$14.5V_mI_m$
I_{R2} $2.2I_m$	$4V_m$	$8.8V_mI_m$	I_{R5} $2.9I_m$	$4V_m$	$11.2V_mI_m$
I_{R3} $3.7I_m$	$3V_m$	$11.1V_mI_m$	I_{R2} $3.1I_m$	$3V_m$	$9.3V_mI_m$
I_{R4} $3.7I_m$	$2V_m$	$7.4V_mI_m$	I_{R3} $3.3I_m$	$2V_m$	$6.6V_mI_m$
I_{R5} $3.7I_m$	$1V_m$	$3.7V_mI_m$	I_{R4} $3.6I_m$	$1V_m$	$3.6V_mI_m$

The current for each row under the configuration of Fig. 2.9(c) is as follows

$$I_{R1} = I_{R5} = 1 \times 0.5I_m + 1 \times 0.2I_m + 2 \times 0.9I_m + 1 \times 0.4I_m = 2.9I_m$$
$$I_{R2} = 1 \times 0.5I_m + 1 \times 0.2I_m + 2 \times 0.9I_m + 1 \times 0.6I_m = 3.1I_m \qquad (2.13)$$
$$I_{R3} = I_{R4} = 1 \times 0.5I_m + 1 \times 0.4I_m + 2 \times 0.9I_m + 1 \times 0.6I_m = 3.3I_m$$

The theoretical electrical values such as current, voltage, and power are taken into account to find the GMPP for conventional and configuration under long wide shading conditions is tabulated in Table 2.5. The Voltage-Current and Voltage-Power characteristic curves are shown in Fig. 2.11. From these curves it is quite evident that two local peaks are developed in these characteristics under Long Wide shading conditions. The configuration minimized two peaks to a single peak and also increased the power yield under same shading conditions.

A small increment in power of 11W is observed for the TCT configuration compared with SP configuration. Moreover, there is no change in power when the configuration is switched from TCT to BL and HC because more than 60% of the array

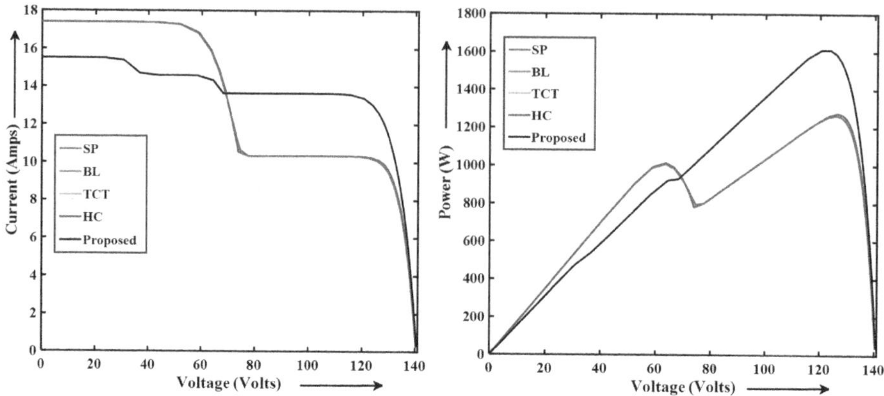

Figure 2.11 Voltage-Current and Voltage-Power characteristics for SP, BL, TCT, HC, and Tom-Tom configurations for Long Wide shading conditions.

Table 2.6

Generated Output Power (W) and Overall Power Loss Caused by Shading for Different Configurations Under Long Wide Shading Condition

Topology	Power (W)	Power Loss (W)	Power Loss (%)
SP	1265	1485	54
BL	1276	1474	53.6
TCT	1276	1474	53.6
HC	1276	1474	53.6
TT	**1613**	**1137**	**41.34**

is shaded and the cross ties don't show that much effect. Nevertheless, power increment of 337W is observed for the Tom-Tom configuration when compared with TCT configuration. The out-put power shows improvement because mismatch current is minimized as shading is dispersed throughout the array by using Tom-Tom configuration. It is observed from Table 2.6 that configuration shows low power loss in comparison with other configurations. The output power produced by the configuration (1613W) is high along with the minimum power loss (1137W, 41.34%) when compared to SP (1485W, 54%), BL (1474W, 53.6%), HC (1474W, 53.6%) and TCT (1474W, 53.6%). The output power yield, and the power losses clearly state that the configuration works efficiently and effectively even when the array is heavily shaded.

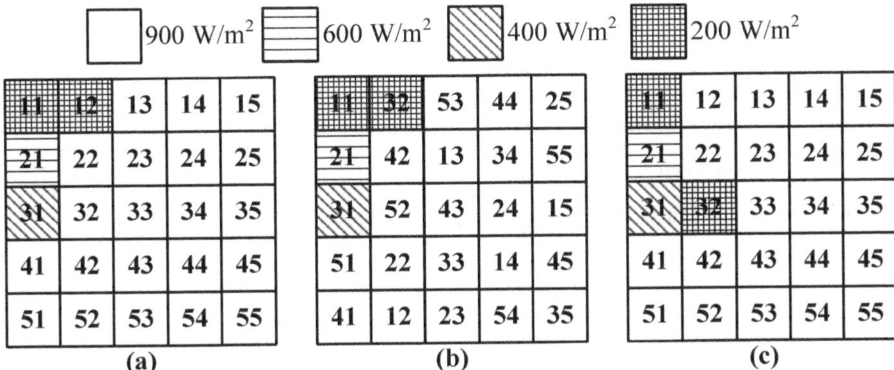

Figure 2.12   Pattern for Long Narrow static shading conditions (a) TCT configuration (b) Tom-Tom configuration (c) Dispersing the shade with Tom-Tom configuration.

2.3.1.4 Long Narrow Shading Conditions

This case is similar to case-1 (Short Narrow) under which almost one-fifth of the panels are shaded. This shading case has minimum shaded panels. The four groups of shadings level are taken $900W/m^2$, $600W/m^2$, $400W/m^2$, and $200W/m^2$. The static shading condition applied for conventional configurations is depicted in Fig. 2.12(a).

The row currents for the TCT configuration are given as

$$\begin{aligned}
I_{R1} &= 2\times0.2I_m + 3\times0.9I_m = 3.1I_m \\
I_{R2} &= 1\times0.6I_m + 4\times0.9I_m = 4.2I_m \\
I_{R3} &= 1\times0.4I_m + 4\times0.9I_m = 4.0I_m \\
I_{R4} &= I_{R5} = 5\times0.9I_m = 4.5I_m
\end{aligned}$$
(2.14)

The row currents for the Tom-Tom configuration are given as

$$\begin{aligned}
I_{R1} &= 1\times0.2I_m + 4\times0.9I_m = 3.8I_m \\
I_{R2} &= 1\times0.6I_m + 4\times0.9I_m = 4.2I_m \\
I_{R3} &= 1\times0.2I_m + 1\times0.4I_m + 3\times0.9I_m = 3.3I_m \\
I_{R4} &= I_{R5} = 5\times0.9I_m = 4.5I_m
\end{aligned}$$
(2.15)

The theoretical electrical values such as current, voltage, and power to find the location of GMPP for conventional and configuration under long narrow shading conditions is tabulated in Table 2.7. The Voltage-Current and Voltage-Power characteristic curves are represented in Fig. 2.13. Though a local peak is achieved, the power yield is enhanced which is shown in Fig. 2.13.

Table 2.8 shows the summary of power generated (W) and power loss (W & %) for different configurations under long narrow shading conditions. A large chunk of power is lost in SP configuration with 36.25% while minimum power loss is recorded

Table 2.7

Maximum Power Point Position in Total Cross Tied and Tom-Tom Configurations for Long Narrow Shade Conditions

Total Cross Tied Configuration			Tom-Tom Configuration		
Current (A)	**Voltage (V)**	**Power (W)**	**Current (A)**	**Voltage (V)**	**Power (W)**
I_{R1} $3.1I_m$	$5V_m$	$15.6V_mI_m$	I_{R3} $3.3I_m$	$5V_m$	$16.5V_mI_m$
I_{R3} $4I_m$	$4V_m$	$16V_mI_m$	I_{R1} $3.8I_m$	$4V_m$	$15.2V_mI_m$
I_{R2} $4.2I_m$	$3V_m$	$12.6V_mI_m$	I_{R2} $4.2I_m$	$3V_m$	$12.6V_mI_m$
I_{R4} $4.5I_m$	$2V_m$	$9V_mI_m$	I_{R4} $4.5I_m$	$2V_m$	$9V_mI_m$
I_{R5} $4.5I_m$	$1V_m$	$4.5V_mI_m$	I_{R5} $4.5I_m$	$1V_m$	$4.5V_mI_m$

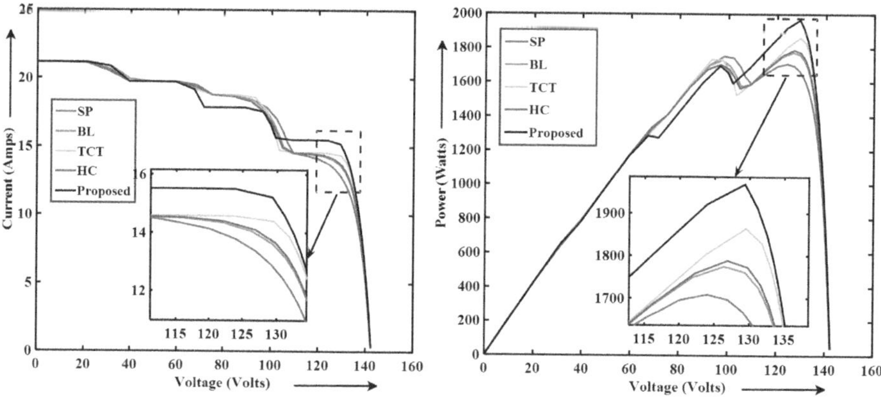

Figure 2.13 Voltage-Current and Voltage-Power characteristics for SP, BL, TCT, HC, and Tom-Tom configurations for Long Wide shading conditions.

up to 28.43% with Tom-Tom configurations owing to the Long Narrow shading conditions. The change in the configuration of the panels leads to dispersion of shading pattern over the entire array. The comparison of different configurations with the Tom-Tom configuration, shows the increment of power for the configuration at 10.9% against SP configuration, 9.8% when compared with BL configuration, 2.3% when compared with HC configuration, and 5% against the TCT configuration.

It is clearly seen that under all non-uniform shading conditions; the configuration possess maximum power output with minimum shading losses than the conventional configurations. The configuration is also tested under progressive incremental moving shading conditions and the results are explained in the next sub-section.

2.3.2 PROGRESSIVE INCREMENTAL MOVING SHADING CONDITIONS

The superiority of the configurations when compared with conventional configurations has been tested under Progressive Incremental Top to Bottom Moving shading

Table 2.8
Generated Output Power (W) and Overall Power Loss Caused by Shading for Different Configurations under Long Narrow Shading Conditions

Topology	Power (W)	Power Loss (W)	Power Loss (%)
SP	1753	997	36.25
BL	1774	976	35.49
TCT	1863	887	32.25
HC	1788	962	34.98
TT	**1968**	**782**	**28.43**

conditions and the performance is detailed below by taking into account the maximum output and Mismatch losses.

These are dynamic shading or moving shading conditions that cover almost five different shadings in each shading condition. The main source for these types of shading conditions is due to neighboring tall erected structures and moving clouds. These shading conditions gives the representation of moving cloud with varying density which is increasing with its movement from initial to final point. The main advantage associated with this shading condition is that it can function both as a dynamic shade condition when considered as a pattern and a static shade condition when individual cases are considered. The shading is taken on all the panels which is incremented progressively in Top to Bottom as shown in Fig. 2.14. The movement of cloud starts from case-1(a) and ends at case-1(e) in each shading condition.

These are dynamic shading or shifting shading conditions that each encompass almost five distinct shadings. The primary cause of these shade situations is due to nearby towering constructed buildings and moving clouds. These shading conditions illustrate a moving cloud with varied density that increases as it moves from the start to the end position. The fundamental benefit of this shading condition is that it may operate as both a dynamic shade condition when examined as a pattern and a static shade condition when considering individual occurrences. As seen in Fig. 2.14, the shade is applied to all of the panels and is increased gradually from top to bottom. In each shading condition, cloud movement begins at case-1(a) and concludes at case-1(e).

2.3.2.1 Voltage-Power and Voltage-Current Characteristics and Maximum Output Power

The electrical characteristics such as voltage, current, and power are computed and tabulated in Table 2.9, and it is clear from the table that power is increased in virtually all of the situations. This increase in power is due to the shadow being diffused by the Tom-Tom setup. The Voltage-Current and Voltage-Power characteristics of

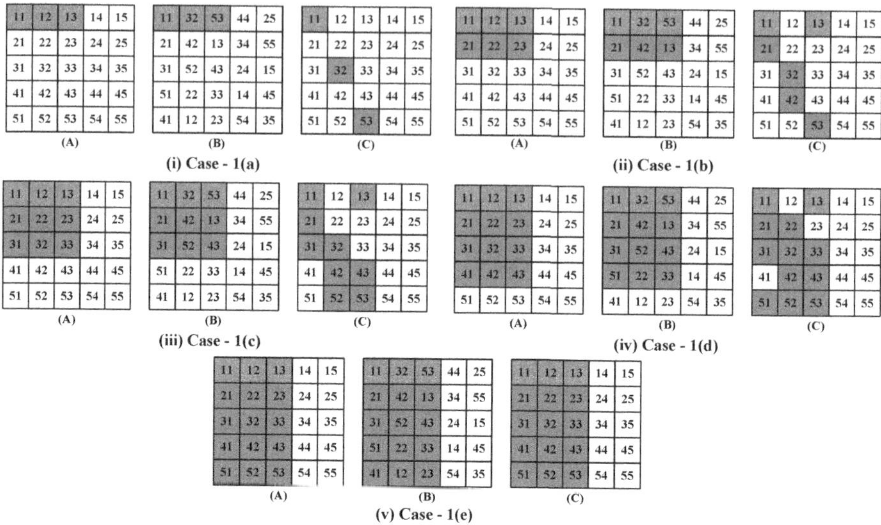

Figure 2.14 Shading pattern for Progressive Incremental Top to Bottom Moving shading conditions (A) Shading on Conventional configuration (B) Shading on Tom-Tom configuration (C) Shade dispersion with configuration.

the SPV Array under Progressive Incremental Top to Bottom Moving shading circumstances are shown in Fig. 2.15 for the conventional and suggested designs. The findings show that the suggested Tom-Tom configuration increases output power in all shading scenarios.

In case-1(a) (Fig. 2.14 (i)), three panels in row-1 (top row) receive an irradiation level of $300W/m^2$, the panels in remaining rows receive an irradiation level of $900W/m^2$. The Voltage-Current and Voltage-Power characteristic curves for different configurations under this type of shading conditions are shown in Fig. 2.15(i). The characteristic curves illustrates that the shade dispersion based on the configuration produces more output power compared to the conventional configurations and the local peaks caused due to the shading are eliminated by the configuration because of shade dispersion. Among all the conventional configurations, TCT possess the maximum output power of 1839W with a power loss of 911W (33.13%), SP delivers an output power of 1858W with a power loss of 892W (32.44%). BL and HC configurations possess approximately equal amount of power with a power loss of 919W (33.42%), 918W (33.38%) respectively. The reconfiguration through Tom-Tom puzzle pattern delivers maximum power of 2188W with a power loss of 562W (20.44%). The summary of all these values are tabulated in Table 2.9. It is clearly visible that the configuration decreases the power loss and increases the output yield.

During the second case in Progressive Incremental Top to Bottom Moving shading conditions (Fig. 2.14(ii)), three panels in each row-1 and row-2 receive an irradiation level of $300W/m^2$, the panels in remaining rows receive an irradiation level of $900W/m^2$. The Voltage-Current and Voltage-Power characteristic curves for

Table 2.9

Position of Maximum Power Point in Total Cross Tied and Tom-Tom Configuration for Progressive Incremental Moving Shading Conditions

Case-1(a)					
Total Cross Tied Configuration			**Tom-Tom Configuration**		
Current(A)	**Voltage(V)**	**Power(W)**	**Current(A)**	**Voltage(V)**	**Power(W)**
I_{R1} $2.7I_m$	$5V_m$	$13.5V_mI_m$	I_{R1} $3.9I_m$	$5V_m$	$19.5V_mI_m$
I_{R2} $4.5I_m$	$4V_m$	$18V_mI_m$	I_{R2} $3.9I_m$	$4V_m$	$15.6V_mI_m$
I_{R3} $4.5I_m$	$3V_m$	$13.5V_mI_m$	I_{R3} $3.1I_m$	$3V_m$	$11.7V_mI_m$
I_{R4} $4.5I_m$	$2V_m$	$9V_mI_m$	I_{R4} $4.5I_m$	$2V_m$	$9V_mI_m$
I_{R5} $4.5I_m$	$1V_m$	$4.5V_mI_m$	I_{R5} $4.5I_m$	$1V_m$	$4.5V_mI_m$
Case-1(b)					
I_{R2} $2.7I_m$	$5V_m$	$13.5V_mI_m$	I_{R1} $3.3I_m$	$5V_m$	$16.5V_mI_m$
I_{R1} $2.7I_m$	$4V_m$	$10.8V_mI_m$	I_{R2} $3.9I_m$	$4V_m$	$15.6V_mI_m$
I_{R3} $4.5I_m$	$3V_m$	$13.5V_mI_m$	I_{R3} $3.9I_m$	$3V_m$	$11.7V_mI_m$
I_{R4} $4.5I_m$	$2V_m$	$9V_mI_m$	I_{R4} $3.9I_m$	$2V_m$	$7.8V_mI_m$
I_{R5} $4.5I_m$	$1V_m$	$4.5V_mI_m$	I_{R5} $3.9I_m$	$1V_m$	$3.9V_mI_m$
Case-1(c)					
I_{R2} $2.7I_m$	$5V_m$	$13.5V_mI_m$	I_{R1} $3.3I_m$	$5V_m$	$16.5V_mI_m$
I_{R1} $2.7I_m$	$4V_m$	$10.8V_mI_m$	I_{R2} $3.3I_m$	$4V_m$	$13.2V_mI_m$
I_{R3} $2.7I_m$	$3V_m$	$8.1V_mI_m$	I_{R3} $3.3I_m$	$3V_m$	$9.9V_mI_m$
I_{R4} $4.5I_m$	$2V_m$	$9V_mI_m$	I_{R4} $3.3I_m$	$2V_m$	$6.6V_mI_m$
I_{R5} $4.5I_m$	$1V_m$	$4.5V_mI_m$	I_{R5} $3.9I_m$	V_m	$3.9V_mI_m$
Case-1(d)					
I_{R2} $2.7I_m$	$5V_m$	$13.5V_mI_m$	I_{R1} $2.7I_m$	$5V_m$	$13.5V_mI_m$
I_{R1} $2.7I_m$	$4V_m$	$10.8V_mI_m$	I_{R2} $2.7I_m$	$4V_m$	$10.8V_mI_m$
I_{R3} $2.7I_m$	$3V_m$	$8.1V_mI_m$	I_{R3} $3.3I_m$	$3V_m$	$9.9V_mI_m$
I_{R4} $2.7I_m$	$2V_m$	$5.4V_mI_m$	I_{R4} $3.3I_m$	$2V_m$	$6.6V_mI_m$
I_{R5} $4.5I_m$	$1V_m$	$4.5V_mI_m$	I_{R5} $3.3I_m$	V_m	$3.3V_mI_m$
Case-1(e)					
I_{R2} $2.7I_m$	$5V_m$	$13.5V_mI_m$	I_{R1} $2.7I_m$	$5V_m$	$13.5V_mI_m$
I_{R1} $2.7I_m$	$4V_m$	$10.8V_mI_m$	I_{R2} $2.7I_m$	$4V_m$	$10.8V_mI_m$
I_{R3} $2.7I_m$	$3V_m$	$8.1V_mI_m$	I_{R3} $2.7I_m$	$3V_m$	$8.1V_mI_m$
I_{R4} $2.7I_m$	$2V_m$	$5.4V_mI_m$	I_{R4} $2.7I_m$	$2V_m$	$5.4V_mI_m$
I_{R5} $2.7I_m$	V_m	$2.7V_mI_m$	I_{R5} $2.7I_m$	V_m	$2.7V_mI_m$

different configurations under this type of shading conditions are shown in Fig. 2.15(ii). The chronological order increment in output power produced for all the configurations is 1427W, 1505W, 1538W, 1580W, and 1741W for HC, SP, BL, TCT, and Tom-Tom based reconfiguration, respectively. The generated power along with its power loss is depicted thoroughly in Table 2.10.

In this case, three panels in first three rows are shaded and remaining two are unshaded. The irradiance profile is shown in Fig. 2.14(iii). The Voltage-Current and Voltage-Power characteristic curves for different configurations under this type of shading conditions are shown in Fig. 2.15(iii). The power developed by the SP configuration is least among all the conventional configurations of 1487W. TCT being most rugged and benchmark configuration, possess maximum power enhancement and the output yield is attained to be 1526W. In addition, the output power obtained

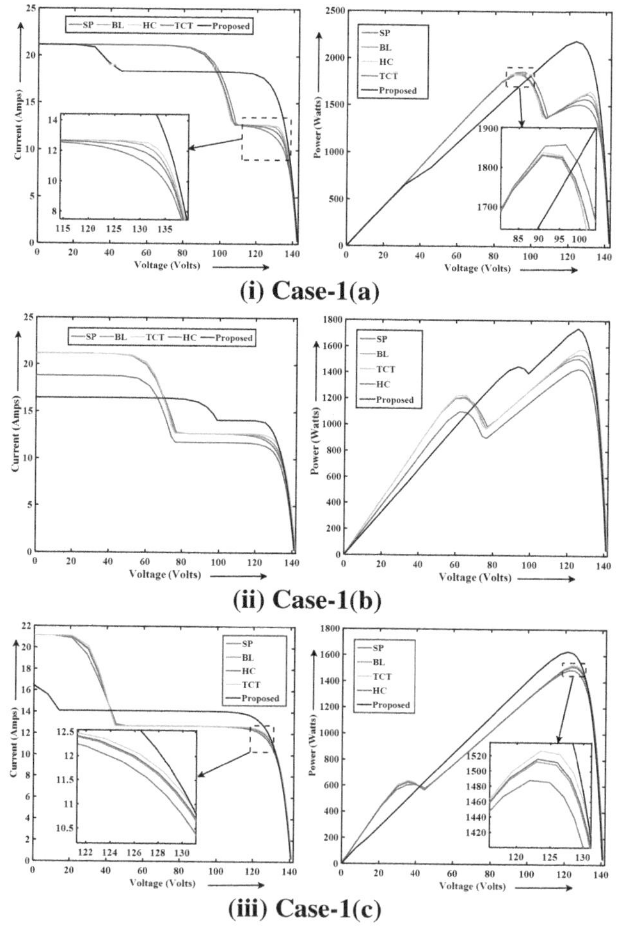

(i) Case-1(a)

(ii) Case-1(b)

(iii) Case-1(c)

Figure 2.15 Voltage-Current and Voltage-Power characteristics for SP, BL, TCT, HC, and Tom-Tom configurations for incremental moving shading conditions.

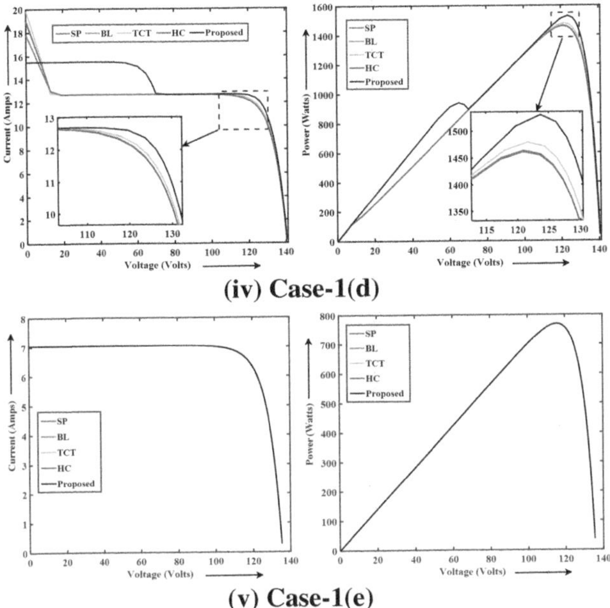

(iv) Case-1(d)

(v) Case-1(e)

Figure 2.15 Continued.

by the configuration is sufficiently more than the conventional configurations with the least power loss which is depicted in Table 2.10.

As aforementioned, the case-1(d) is also the progressive increment in row wise manner than the previous case-(c). The moving clouds is considered to cover three panels in all the four rows in the array, i.e., the panels in row-1 to row-4 are shaded with $300W/m^2$ and unshaded row with $900W/m^2$. The shading profile along with dispersion is shown in Fig. 2.14(iv). It is observed that SP, BL, TCT, and HC generates the output power of 1463W, 1461W, 1479W, and 1461W, respectively. BL and HC possess the same output power because of the number of interconnections between them are almost same. Tom-Tom reconfiguration generates 1530W, which is 51W more than the conventional configuration clearly tabulated in Table 2.10. The Voltage-Current and Voltage-Power characteristic curves for different configurations under this type of shading conditions are shown in Fig. 2.15(iv).

The cloud has started from its initial position in case-(a) and came to final position in this case-(e). The shading profile is shown in Fig. 2.14(v). As the entire array is shaded, the output power is same in all configurations is recorded at 1434W. The Voltage-Current and Voltage-Power characteristic curves are plotted in Fig. 2.15(v) and the values are tabulated in Table 2.10.

2.3.2.2 Mismatch Power Loss From Normal Conditions to the Shading Conditions

The suggested Progressive Incremental Top to Bottom configuration: When compared to the other configurations for the relevant shading situation, moving shading

Table 2.10

Generated Power (W) and Power Loss Caused by Shading for Different Topologies under Progressive Incremental Top to Bottom Moving Shading Conditions

	Topology	Power (W)	Power Loss (W)	Power Loss (%)
Case – 1(a)	SP	1858	892	32.44
	BL	1831	919	33.42
	TCT	1839	911	33.13
	HC	1832	918	33.38
	TT	2188	562	20.44
Case – 1(b)	SP	1505	1245	45.27
	BL	1538	1212	44.07
	TCT	1580	1170	42.55
	HC	1427	1323	48.11
	TT	**1741**	**1009**	**36.69**
Case – 1(c)	SP	1487	1263	45.93
	BL	1512	1238	45.02
	TCT	1526	1224	44.51
	HC	1516	1234	44.87
	TT	**1630**	**1120**	**40.73**
Case – 1(d)	SP	1463	1287	46.8
	BL	1461	1289	46.87
	TCT	1479	1271	46.22
	HC	1461	1289	46.87
	TT	**1530**	**1220**	**44.36**
Case – 1(e)	SP	1434	1316	47.85
	BL	1434	1316	47.85
	TCT	1434	1316	47.85
	HC	1434	1316	47.85
	TT	**1434**	**1316**	**47.85**

conditions result in much reduced mismatch power loss (Table 2.10). A maximum of 10% mismatch power loss is reported for case-1(c), when one-fourth of the panels are shaded under these circumstances.

The maximum power obtained and its mismatch power loss for progressive incremental diagonal moving shading conditions (Fig. 2.16) and left to right (Fig. 2.17) shading conditions are given in Table 2.11 and Table 2.12. From the tables, it can be seen that the method is giving better results than all the existing methods for all the moving and static shading conditions.

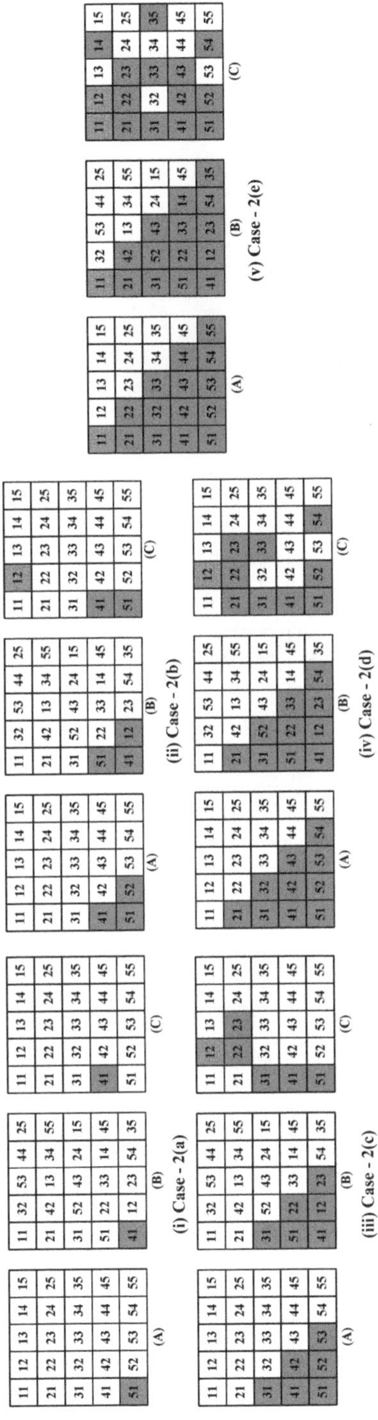

Figure 2.16 Shading pattern for Progressive Incremental Diagonal Moving shading conditions (A) Shading on Conventional configuration (B) Shading on Tom-Tom configuration (C) Shade dispersion with configuration.

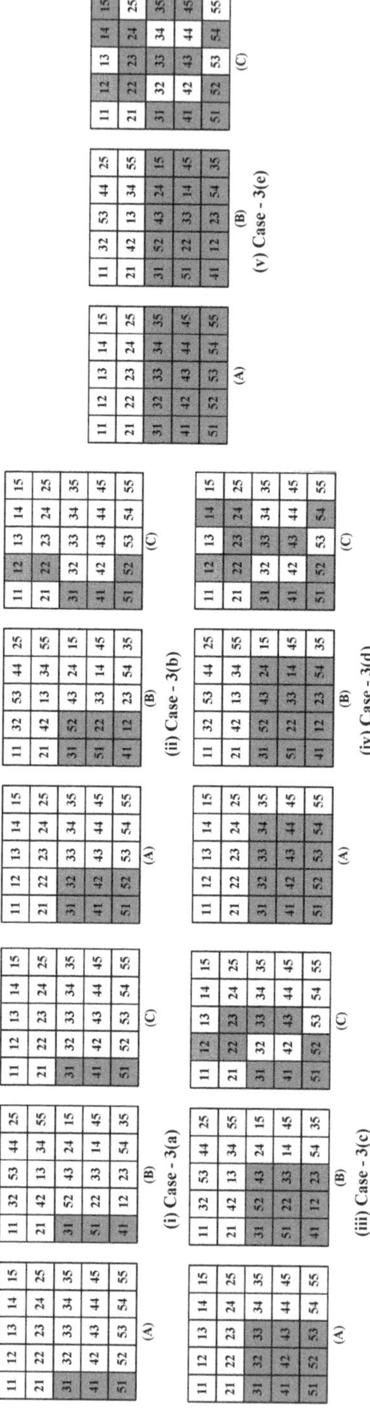

Figure 2.17 Shading pattern for Progressive Incremental Left to right Moving shading conditions (A) Shading on Conventional configuration (B) Shading on Tom-Tom configuration (C) Shade dispersion with configuration.

Table 2.11

Generated Output Power (W) and Power Loss Caused by Shading for Different Topologies under Progressive Incremental Diagonal Moving Shading Conditions

	Maximum Output Power (W)					Power Losses (W)				
	SP	BL	TCT	HC	TT	SP	BL	TCT	HC	TT
Case-2(a)	2141	2194	2291	2235	**2291**	609	556	459	515	**459**
Case-2(b)	1824	1920	1976	1898	**2188**	926	830	774	852	**562**
Case-2(c)	1499	1567	1621	1544	**1929**	1251	1183	1129	1206	**821**
Case-2(d)	1100	1161	1180	1145	**1506**	1650	1589	1570	1605	**1244**
Case-2(e)	891	933	946	919	**1434**	1859	1817	1804	1831	**1316**

Table 2.12

Generated Output Power (W) and Power Loss Caused by Shading for Different Topologies under Progressive Incremental Left to Right Moving Shading Conditions

	Maximum Output Power (W)					Power Losses (W)				
	SP	BL	TCT	HC	TT	SP	BL	TCT	HC	TT
Case-3(a)	2136	2142	2188	2144	**2188**	614	608	562	606	**562**
Case-3(b)	1811	1835	1861	1918	**1929**	939	915	889	832	**821**
Case-3(c)	1487	1493	1526	1466	**1800**	1263	1257	1224	1284	**950**
Case-3(d)	1165	1183	1185	1005	**1530**	1585	1567	1565	1745	**1220**
Case-3(e)	841	841	841	841	**1434**	1909	1909	1909	1909	**1316**

2.4 SUMMARY

Partial shading of panels in SPV array is one of the major drawbacks of the solar PV system which leads to substantial decrement in the output power. Additionally, this also disrupt the characteristic curves by causing in multiple maximum peaks. This problem can be solved by adopting efficient reconfiguration of panels. In this chapter, a novel Tom-Tom configuration is used in order to mitigate the effects of shading. The reconfiguration follows static configuration under which the physical location of the shaded and unshaded panels is modified without changing the electrical circuitry in order to minimize the effects of shading and also to reduce mismatch of row currents. This reconfiguration is tested under four different non-uniform shading conditions and progressive incremental moving shading conditions (equivalent to a cloud which is moving from initial to final position). The results prove enhanced improvement in the maximum output power against the existing conventional configurations for a 5×5 SPV array for various shading scenarios with minimum power loss. The major disadvantage with this Tom-Tom reconfiguration method is that this method is applicable to only 5×5 SPV array and it is not applicable to different sizes.

3 Array Reconfiguration with Arrow Sudoku Puzzle Pattern

3.1 INTRODUCTION

As evident from the previous chapter that Tom-Tom approach is applicable only to 5×5 SPV Array which limits its scalability and applicability. Hence, a new algorithm for mitigating the effects of partial shading is developed and implemented for a 6×6 SPV Array. This algorithm is derived from the family of Sudoku puzzles named as Arrow Sudoku Puzzle pattern. The reconfiguration is proved to be the potential topology because of the following features.

- The proposed reconfiguration based on Arrow Sudoku puzzle helps to distribute the shading pattern throughout the array which leads to reduction of mismatch losses in the row currents and also enhance the output power.
- The dispersion of the shading pattern based on proposed configuration yields improved output power against all conventional and hybrid configurations because the proposed configurations help to better balance the row currents reducing the mismatch losses which is not possible with other configurations.

The rest of the chapter is organized as follows: Section 3.2 describes the proposed approach for array reconfiguration. Section 3.3 discusses about the results when compared with the conventional configurations. Finally, the chapter is summarized in Section 3.4

3.2 PROPOSED ARROW SUDOKU SHADE DISPERSION RECONFIGURATION TOPOLOGY

All though the TCT interconnection gives improved output power but it suffers from the drawback of zero shade dispersion which causes multiple peaks in the characteristics curve. The electrical interconnection limits the ability of TCT configuration to distribute the shading pattern across the array grid. Consequently, the row current, which is dependent upon the irradiance levels, swings across a wide range of values. So, the shaded panels which are generating a low level of row current are bypassed to insulate them from being getting damaged. Therefore, in TCT method, bypassing of the shaded panels leads to stepped voltage-current characteristics and thus expanded mismatch losses appear in the system. Hence, in order to better utilize the shaded panels rather than bypassing them can be done by rearrangement of the

DOI: 10.1201/9781003285830-3

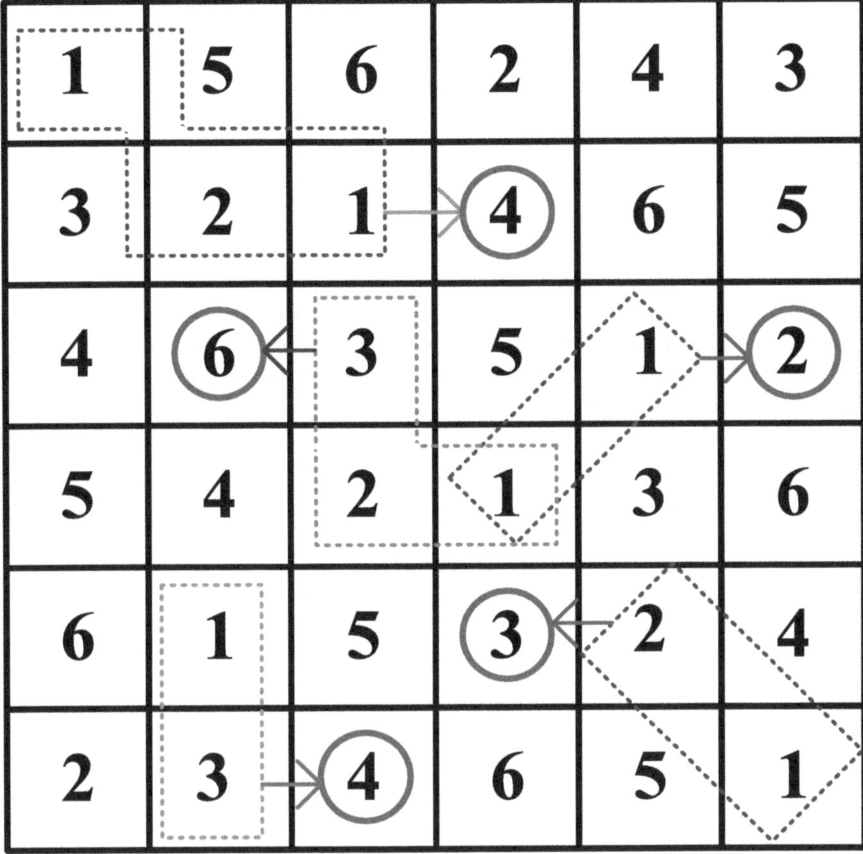

Figure 3.1 Proposed Arrow Sudoku puzzle pattern.

shaded panels across the array which will ultimately improve the output power for the TCT configuration along with reduction in mismatch losses. A novel arrangement, Arrow Sudoku, is offered to disperse the shadow throughout the full array to lessen the impact of shading.

Arrow Sudoku is an advanced form of Sudoku that requires more precision and accuracy. All of the empty squares in the original 6×6 Sudoku are filled with the digits 1 to 6 such that they occur precisely once in each row, column, and 3-by-2 box (which we may call as a region). The Arrow Sudoku, a modification of the original Sudoku, is comparable to the latter with the exception of the use of sub-grids and arrows. The sub-grids with numbers are located at one end of the arrow (tail end). The opposite end of the arrow (the head end) is pointing at a ringed number. The traditional Sudoku principles may be used to solve the arrow Sudoku problem, and the total of the numbers in the sub-grids of the tail end arrow must equal the arrow-headed encircling number. For example (Fig. 3.1), if the sub-grid includes the

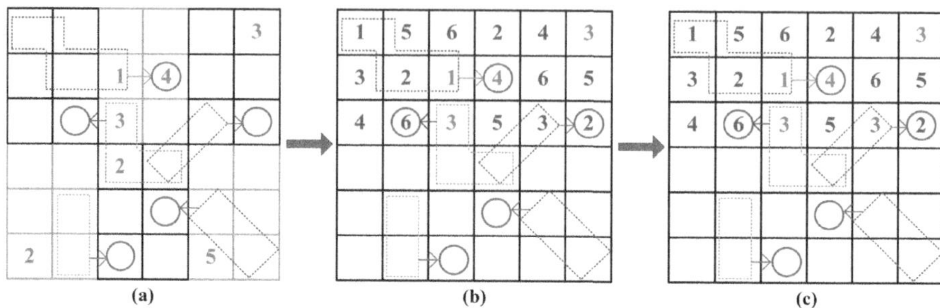

Figure 3.2 (a) Proposed array with mathematical operations on the left upper corner of each sub-grid. (b) Filling the vacant cells with integers ranging from 1 to 5. (indicated in black color) (c) Failure of the suggested Arrow Sudoku puzzle design criterion (indicated in Red Color).

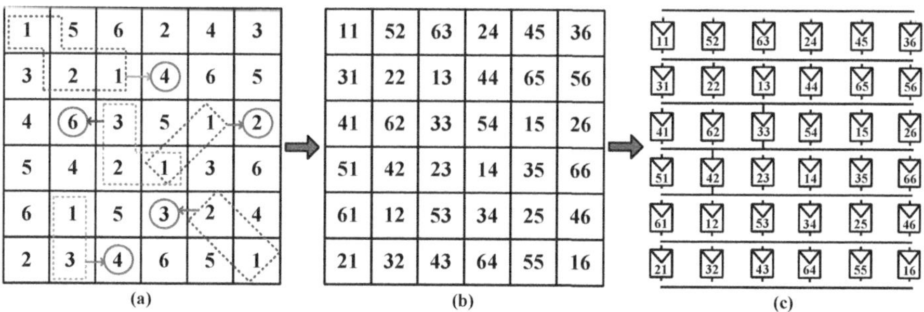

Figure 3.3 (a) Fully completed puzzle pattern (b) Final Arrow Sudoku reconfiguration with column numbers used for the study (c) Initial unconnected renumbering of panels in the array.

numbers 1, 2, and 3, the circled number must be the total of 1, 2, and 3, which equals 6. A number may be repeated in the sub-grids followed by an arrow if none of the original Sudoku rules are broken, i.e., the sub-grid can include the same number only if it does not belong to the same row or column or that specific 3-by-2 section where the repetition occurs. As a result, if an arrow crosses many zones, it may include 1, 1 which total to 2, as illustrated in Fig. 3.1. Because of the inclusion of arrows, Sudoku logic alone will not enough to solve the Arrow Sudoku. All of the numbers in the circle, as well as those in the sub-grid, must be searched. Fluke logic's would not operate in this perplexing pattern since it has a unique solution that can be obtained simply by doing an addition operation. The process of solving or completing the Arrow Sudoku Puzzle Pattern is discussed in Fig. 3.2, and the final pattern is displayed in Fig. 3.3(a).

The numbers present in the Arrow Sudoku puzzle indicates the row number and for the column number an extra number is added to generate a 6×6 SPV Array i.e., in Fig. 3.3(b), number 6 present in the second column of third row is coined as '62'. The

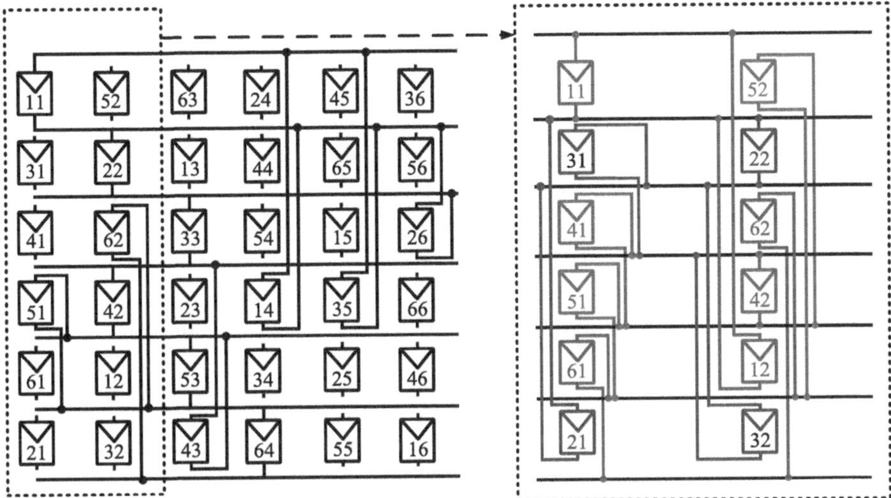

Figure 3.4 (a) Final Arrow Sudoku reconfiguration (b) Enlarged version with all connections for first two columns of (a).

main benefit associated with Arrow Sudoku is that it distributes the shading pattern assembled in one spot in the array over to the entire array. In order to distribute the shading pattern uniformly and efficiently throughout the entire array, panel having the same number should not be on either side of that panel.

The first renumbering of the panels in the array is shown in Fig. 3.3(c), which does not demonstrate any connections between the numbers. Fig. 3.4(a) depicts the structural arrangement, and Fig. 3.4(b) is an expanded version of Fig. 3.4(a) that shows all of the connections for two columns (b).

Here, relocation of the panels needs to be done only once that to involve only physical re-arrangement without disturbing the electrical connections. So, this means that the current and voltage equations used for TCT configuration can also be used for rearranged configuration. The panels present in a particular row for the TCT configuration is now moved to other location within the array. For example, the panel with the number 35, which was previously located in the third row and fifth column of the TCT arrangement, has been moved to the fourth row and fifth column of the proposed configuration. However, its electrical connection remains the same as it was in the third row and second column of the TCT arrangement, as can be seen in Fig. 3.4 (b).

The advantages of the proposed Arrow Sudoku shade dispersion method are as follows:

1. The main objective of the Arrow Sudoku is to distribute the shaded panels gathered at a particular spot in the array over to the entire array. In order to distribute the shading pattern effectively and efficiently, panel having the same

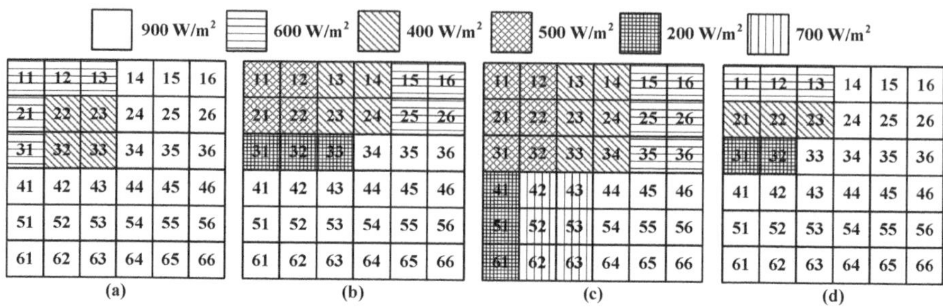

Figure 3.5   Shading profile for 6×6 under Non-Uniform shading conditions used in this work.

number should not be present on the either side of that panel in any row, which can be clearly observed in Fig. 3.4(a) and Fig. 3.4(b).

2. The proposed configuration has an added advantage of economically viable as it does not require any extra connections.

3. The goal of the proposed configuration is not only to enhance the performance parameters of the array but also to execute the configuration on a large-scale array.

3.3   RESULTS AND DISCUSSIONS

To validate the ability of the proposed Arrow Sudoku reconfiguration against different types of interconnections under various non-uniform and incremental moving shading conditions by taking into account the parameters like Maximum output power and power loss are done in the subsequent subsections along with the analyzes of the results.

3.3.1   ANALYSIS OF THE RESULTS FOR THE NON-UNIFORM SHADING CONDITIONS

Four distinct non-uniform shading conditions such as Short Narrow, Short Wide, Long Wide, and Long Narrow are used in this work. For 6×6 array, each of the non-uniform shading conditions mentioned are depicted in Fig. 3.5.

3.3.1.1   Short Narrow Shading Conditions

Three different levels of Solar irradiation are taken for this shading condition. Group one is irradiated $900\text{W}/\text{m}^2$, while group two gets $600\text{W}/\text{m}^2$, and group three receives $400\text{W}/\text{m}^2$ as shown in Fig. 3.6. In order to locate the Maximum Power Point, the main requirement is to calculate current for each row.

The current in row-1 is as follows

$$I_{R1} = x_{11}I_{11} + x_{12}I_{12} + x_{13}I_{13} + x_{14}I_{14} + x_{15}I_{15} + x_{16}I_{16} \tag{3.1}$$

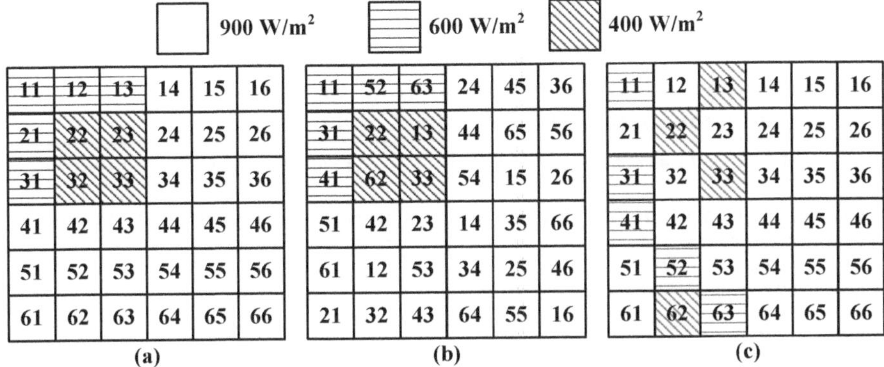

Figure 3.6 Shading Pattern for Short Narrow shading conditions: (a) Conventional configuration; (b) Proposed Arrow Sudoku configuration; (c) Dispersing the shade on to the array with proposed configuration.

Where, $x_{11} = (G_{11}/G_0)$; G_{11} represents the irradiance level received by panel 11. Substituting all the values, the current for row-1 is given as:

$$I_{R1} = 3 \times 0.6 I_m + 3 \times 0.9 I_m = 4.5 I_m \tag{3.2}$$

In row 2 and row 3, one panel each is subjected to 600W/m^2, next two with 400W/m^2 and the other three with 900W/m^2. The current across row 2 is calculated as

$$I_{R2} = I_{R3} = 2 \times 0.4 I_m + 1 \times 0.6 I_m + 3 \times 0.9 I_m = 4.1 I_m \tag{3.3}$$

Rows 4, 5, and 6 are subjected to no shade, the current is given by

$$I_{R4} = I_{R5} = I_{R6} = 6 \times 0.9 I_m = 5.4 I_m \tag{3.4}$$

For proposed Arrow Sudoku configuration shown in Fig. 3.6(c), because of the shade dispersion, the panels are shaded into different rows. So, the currents are given by:

For row 1, only one panel is shaded with 600W/m^2, 400W/m^2 while the remaining all panels are exposed to an irradiation level of 900W/m^2.

$$I_{R1} = 1 \times 0.6 I_m + 1 \times 0.4 I_m + 4 \times 0.9 I_m = 4.6 I_m \tag{3.5}$$

Rows 1, 3, and 6 receive same irradiation levels. So, the currents in all these rows will be same

$$I_{R1} = I_{R3} = I_{R6} = 4.6 I_m \tag{3.6}$$

In Row 2 for proposed configuration, one panel receives an irradiance level of 400W/m^2 and remaining panels are subjected to 900W/m^2. The current equation is given as

$$I_{R2} = 1 \times 0.4 I_m + 5 \times 0.9 I_m = 4.9 I_m \tag{3.7}$$

Table 3.1

Position of Maximum Power Point in Total Cross Tied and Arrow Sudoku Configuration for Short Narrow Shading Conditions

Total Cross Tied Configuration				Arrow Sudoku Configuration			
	Current(A)	Voltage(V)	Power(W)		Current(A)	Voltage(V)	Power(W)
I_{R2}	$4.1I_m$	$6V_m$	$24.6V_mI_m$	I_{R1}	$4.6I_m$	$6V_m$	$27.6V_mI_m$
I_{R3}	$4.1I_m$	$5V_m$	$20.5V_mI_m$	I_{R3}	$4.6I_m$	$5V_m$	$23V_mI_m$
I_{R1}	$4.5I_m$	$4V_m$	$18V_mI_m$	I_{R6}	$4.6I_m$	$4V_m$	$18.4V_mI_m$
I_{R4}	$5.4I_m$	$3V_m$	$16.2V_mI_m$	I_{R2}	$4.9I_m$	$3V_m$	$14.7V_mI_m$
I_{R5}	$5.4I_m$	$2V_m$	$10.8V_mI_m$	I_{R4}	$5.1I_m$	$2V_m$	$10.2V_mI_m$
I_{R6}	$5.4I_m$	V_m	$5.4V_mI_m$	I_{R5}	$5.1I_m$	V_m	$5.1V_mI_m$

In Row 4, one panel is subjected to $600\text{W}/\text{m}^2$ and remaining panels are subjected to $900\text{W}/\text{m}^2$. The current equation is given as.

$$I_{R4} = 1 \times 0.6I_m + 5 \times 0.9I_m = 5.1I_m \tag{3.8}$$

Rows 4 and 5 gets same irradiation levels. So, the current in Row 4 is same as that of Row 5.

$$I_{R4} = I_{R5} = 5.1I_m \tag{3.9}$$

The theoretical values of Currents (under the sequence the panels are bypassed), Voltages, and Powers required to locate the GMPP for conventional and proposed configuration for each and every row under short narrow shading conditions is tabulated in Table 3.1. The Voltage-Current and Voltage-Power characteristics for all conventional and proposed configurations are shown in Fig. 3.7. It is evident for the Short Narrow condition that the proposed configuration based on Arrow Sudoku results in enhanced output power compared with the conventional TCT configuration with no local Peaks, achieving Global Maximum only once. The power enhancement is the result of improved shade dispersion throughout the array.

The results achieved for generated output power (W), power loss (W) and power loss (%) for SP, BL, TCT, HC, and Proposed Arrow Sudoku configurations under Short Narrow shading conditions are given in Table 3.2. Under no shading i.e., at Standard Test Conditions, 6×6 array yields an output power of 3960W. SP yields an output power of 2739W with an overall shading loss of 1221W (30.83%). BL and HC configurations yield an output power of 2808W and 2806W, respectively with a power loss of 1152W (29.09%) and 1154W (29.14%). The mostly used benchmark configuration TCT yields the power of 2857W with a shading loss of 1103W (27.85%), whereas in the proposed Arrow Sudoku configuration, the power yield has increased to 3087W thereby, minimizing the power loss to 873W (22.05%). Hence, it is validated that the proposed configuration is able to give superior results for various shading conditions.

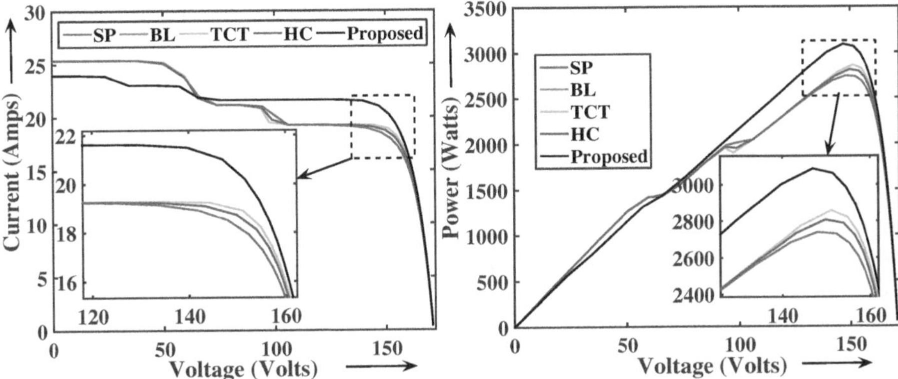

Figure 3.7 Voltage-Current and Voltage-Power characteristics for SP, BL, TCT, HC, and Proposed Arrow Sudoku configurations for Short Narrow shading conditions.

Table 3.2

Generated Output Power (W) and Overall Power Loss Caused by Shading for Different Configurations under Short Narrow Shading Condition

Topology	Power (W)	Power Loss (W)	Power Loss (%)
SP	2739	1221	30.83
BL	2808	1152	29.09
TCT	2857	1103	27.85
HC	2806	1154	29.14
Proposed (AS)	**3087**	**873**	**22.05**

3.3.1.2 Short Wide Shading Conditions

For this shading condition, five divergent levels of irradiations are applied to the Solar PV Array, which are, $900\text{W}/\text{m}^2$, $600\text{W}/\text{m}^2$, $500\text{W}/\text{m}^2$, $400\text{W}/\text{m}^2$, and $200\text{W}/\text{m}^2$. The shading pattern is illustrated in Fig. 3.8.

The current calculations for each row under TCT configuration is as follows

$$I_{R1} = I_{R2} = 2 \times 0.5I_m + 2 \times 0.4I_m + 3 \times 0.6I_m = 3.6I_m$$
$$I_{R3} = 3 \times 0.2I_m + 3 \times 0.9I_m = 3.3I_m \qquad (3.10)$$
$$I_{R4} = I_{R5} = I_{R6} = 6 \times 0.9I_m = 5.4I_m$$

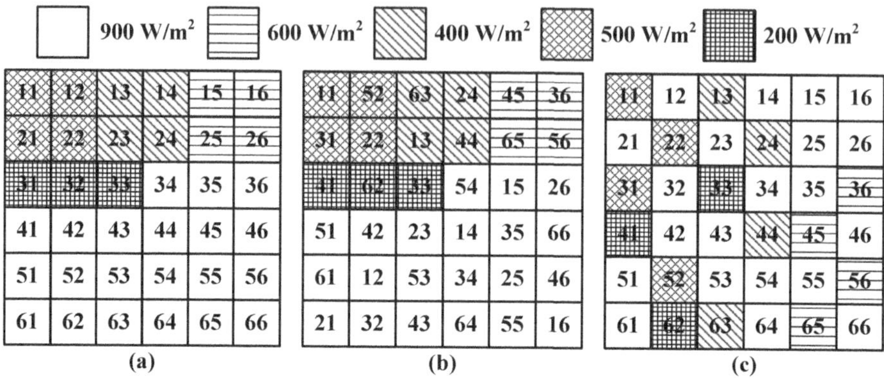

Figure 3.8 Shading pattern for Short Wide (a) Conventional TCT configuration (b) Proposed Arrow Sudoku configuration (c) Shade dispersion with proposed configuration.

The current calculations for each row under the proposed Arrow Sudoku configuration is as follows

$$\begin{aligned}
I_{R1} &= I_{R2} = 1 \times 0.5I_m + 1 \times 0.4I_m + 4 \times 0.9I_m = 4.5I_m \\
I_{R3} &= 1 \times 0.5I_m + 1 \times 0.2I_m + 1 \times 0.6I_m + 3 \times 0.9I_m = 4I_m \\
I_{R4} &= I_{R6} = 1 \times 0.2I_m + 1 \times 0.4I_m + 1 \times 0.6I_m + 3 \times 0.9I_m = 3.9I_m \\
I_{R5} &= 1 \times 0.5I_m + 1 \times 0.6I_m + 4 \times 0.9I_m = 4.7I_m
\end{aligned} \quad (3.11)$$

The theoretical values for Currents (under the sequence the panels are bypassed), Voltages and Powers are employed to get the GMPP for conventional and proposed configuration for each and every row under short wide shading conditions is listed in Table 3.3. The Voltage-Current and Voltage-Power characteristics for all conventional and proposed configurations are shown in Fig. 3.9. A clear observation is made that the disturbances caused by conventional configurations is reduced using the proposed configuration. The output power shows an enhancement of approximately 20% when the array configuration is changed from the TCT configuration to the proposed Arrow Sudoku configuration because of the reshuffling of the panels within the array which results in better shade dispersion across the system.

An incremental increase in power of 412W is obtained when the configuration changes from SP to TCT. An increment in power of 369W is observed when the configuration changes from BL to TCT. An increment in power of 314W is seen when the panels configured as HC changed to TCT configuration. When the configuration is changed from TCT to proposed Arrow Sudoku configuration, an increase in the output power is observed to 495W. The output power shows enhanced increment due to reduction in mismatch currents which is the result of shade dispersion by Arrow Sudoku configurations. It is evident from Table 3.4 that the proposed configuration gives lower power loss compared to other configurations. The output power produced by employing the proposed configuration for the array (2600W) is high with

Table 3.3

Position of Maximum Power Point in Total Cross Tied and Arrow Sudoku Configuration for Short Wide Shading Conditions

Total Cross Tied Configuration			Arrow Sudoku Configuration		
Current(A)	**Voltage(V)**	**Power(W)**	**Current(A)**	**Voltage(V)**	**Power(W)**
I_{R3} $3.3I_m$	$6V_m$	$19.8V_mI_m$	I_{R4} $3.9I_m$	$6V_m$	$23.4V_mI_m$
I_{R1} $3.6I_m$	$5V_m$	$18V_mI_m$	I_{R6} $3.9I_m$	$5V_m$	$19.5V_mI_m$
I_{R2} $3.6I_m$	$4V_m$	$14.4V_mI_m$	I_{R3} $4I_m$	$4V_m$	$16V_mI_m$
I_{R4} $5.4I_m$	$3V_m$	$16.2V_mI_m$	I_{R1} $4.5I_m$	$3V_m$	$13.5V_mI_m$
I_{R5} $5.4I_m$	$2V_m$	$10.8V_mI_m$	I_{R2} $4.5I_m$	$2V_m$	$9V_mI_m$
I_{R6} $5.4I_m$	V_m	$5.4V_mI_m$	I_{R5} $4.7I_m$	V_m	$4.7V_mI_m$

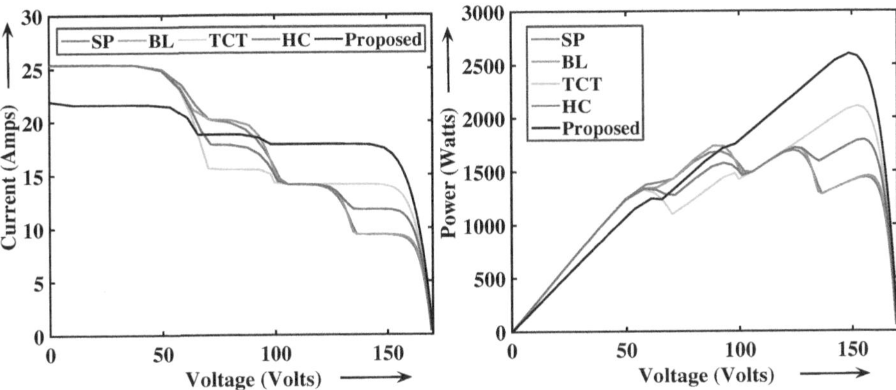

Figure 3.9 Voltage-Current and Voltage-Power characteristics for SP, BL, TCT, HC, and Proposed Arrow Sudoku configurations for Short Wide shading conditions.

the minimum power loss (1360W, 34.34%) when compared to SP (2267W, 57.25%), BL (2224W, 56.16%), HC (2169W, 54.77%), and TCT (1855W, 46.84%). The output power along with the power losses clearly state that the proposed configuration works efficiently and effectively even when the array is heavily shaded.

3.3.1.3 Long Wide Shading Conditions

Under this shading condition most of panels are irradiated with various illumination levels as $900W/m^2$, $750W/m^2$, $600W/m^2$, $500W/m^2$, $400W/m^2$, and $200W/m^2$. The shading pattern is shown in Fig. 3.10. In this type of shading, 70% of panels are shaded in this type of shading condition.

Table 3.4

Generated Output Power (W) and Overall Power Loss Caused by Shading for Different Configurations under Short Wide Shading Condition

Topology	Power (W)	Power Loss (W)	Power Loss (%)
SP	1693	2267	57.25
BL	1736	2224	56.16
TCT	2105	1855	46.84
HC	1791	2169	54.77
Proposed (AS)	**2600**	**1360**	**34.34**

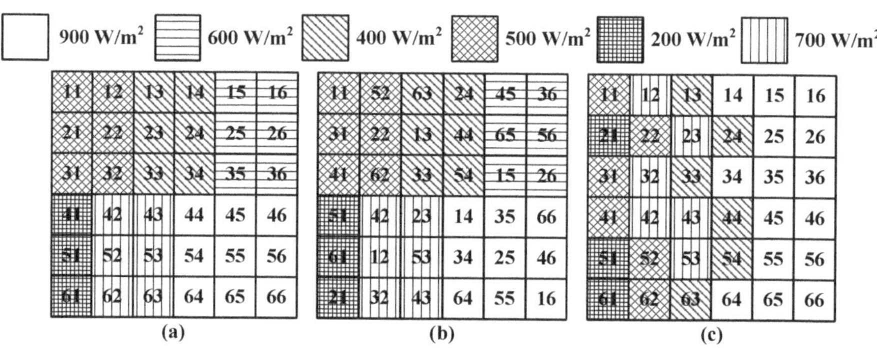

Figure 3.10 Shading pattern for Long Wide (a) Conventional TCT configuration (b) Proposed Arrow Sudoku configuration (c) Shade dispersion with proposed configuration.

The theoretical value of current for each row for the TCT configuration is given as

$$I_{R1} = I_{R2} = I_{R3} = 2 \times 0.6I_m + 2 \times 0.4I_m + 2 \times 0.5I_m = 3I_m$$
$$I_{R4} = I_{R5} = I_{R6} = 1 \times 0.2I_m + 2 \times 0.75I_m + 3 \times 0.9I_m = 4.4I_m \tag{3.12}$$

The current calculations for each row for the proposed Arrow Sudoku configuration are listed as

$$I_{R1} = I_{R3} = 1 \times 0.5I_m + 1 \times 0.75I_m + 1 \times 0.4I_m + 1 \times 0.6I_m + 2 \times 0.9I_m = 4.05I_m$$
$$I_{R2} = 1 \times 0.2I_m + 1 \times 0.5I_m + 1 \times 0.75I_m + 1 \times 0.4I_m + 1 \times 0.6I_m + 1 \times 0.9I_m = 3.35I_m$$
$$I_{R5} = 1 \times 0.2I_m + 1 \times 0.5I_m + 1 \times 0.75I_m + 1 \times 0.4I_m + 1 \times 0.6I_m + 1 \times 0.9I_m = 3.35I_m$$
$$I_{R4} = 1 \times 0.5I_m + 2 \times 0.75I_m + 1 \times 0.4I_m + 1 \times 0.6I_m + 2 \times 0.9I_m = 3.9I_m$$
$$I_{R6} = 1 \times 0.2I_m + 1 \times 0.5I_m + 1 \times 0.4I_m + 1 \times 0.6I_m + 2 \times 0.9I_m = 3.5I_m$$

$$\tag{3.13}$$

Table 3.5

Position of Maximum Power Point in Total Cross Tied and Arrow Sudoku configuration for Long Wide Shading Conditions

Total Cross Tied Configuration			Arrow Sudoku Configuration		
Current(A)	Voltage(V)	Power(W)	Current(A)	Voltage(V)	Power(W)
I_{R1} $3I_m$	$6V_m$	$18V_mI_m$	I_{R2} $3.35I_m$	$6V_m$	$20.1V_mI_m$
I_{R2} $3I_m$	$5V_m$	$15V_mI_m$	I_{R5} $3.35I_m$	$5V_m$	$16.75V_mI_m$
I_{R3} $3I_m$	$4V_m$	$12V_mI_m$	I_{R6} $3.5I_m$	$4V_m$	$14V_mI_m$
I_{R4} $4.4I_m$	$3V_m$	$13.2V_mI_m$	I_{R4} $3.9I_m$	$3V_m$	$11.7V_mI_m$
I_{R5} $4.4I_m$	$2V_m$	$8.8V_mI_m$	I_{R1} $4.05I_m$	$2V_m$	$8.1V_mI_m$
I_{R6} $4.4I_m$	V_m	$4.4V_mI_m$	I_{R3} $4.05I_m$	V_m	$4.05V_mI_m$

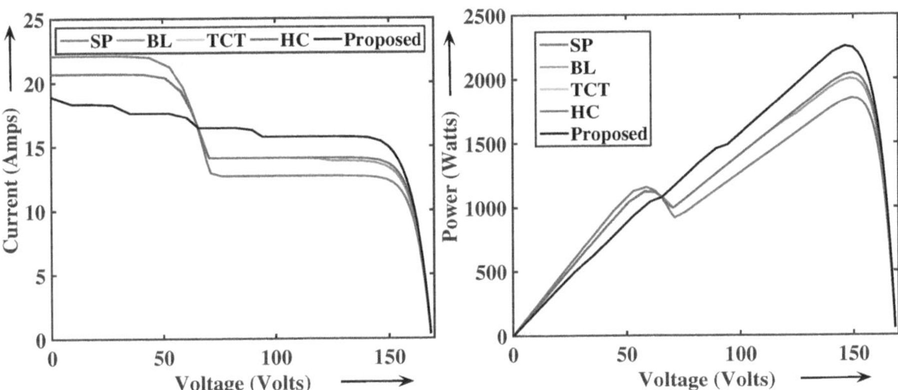

Figure 3.11 Voltage-Current and Voltage-Power characteristics for SP, BL, TCT, HC, and Proposed Arrow Sudoku configurations for Long Wide shading conditions.

The theoretical values for Currents (under the sequence the panels are bypassed), Voltages and Powers are used to locate the GMPP for conventional and proposed configuration for each and every row under long wide shading conditions is tabulated in Table 3.5, and Fig. 3.11 shows the Voltage-Current and Voltage-Power characteristics for all conventional and proposed configurations, it is observed from the figure that the disturbances caused in the conventional configurations is nullified with the proposed configuration.

Table 3.6 shows the summary of power generated (W) and power loss (W & %) for different configurations under long wide shading conditions. A large decrement in power is obtained for SP configuration with 53.33% is contrast with minimum power loss of 43.03% with proposed Arrow Sudoku configurations owing to the Long Wide shading conditions. The configuration of panels helps to mitigate the effects of

Table 3.6

Generated Output Power (W) and Overall Power Loss Caused by Shading for Different Configurations under Long Wide Shading Condition

Topology	Power (W)	Power Loss (W)	Power Loss (%)
SP	1848	2112	53.33
BL	2002	1958	49.44
TCT	2044	1916	48.38
HC	1860	2100	53.03
Proposed (AS)	**2256**	**1704**	**43.03**

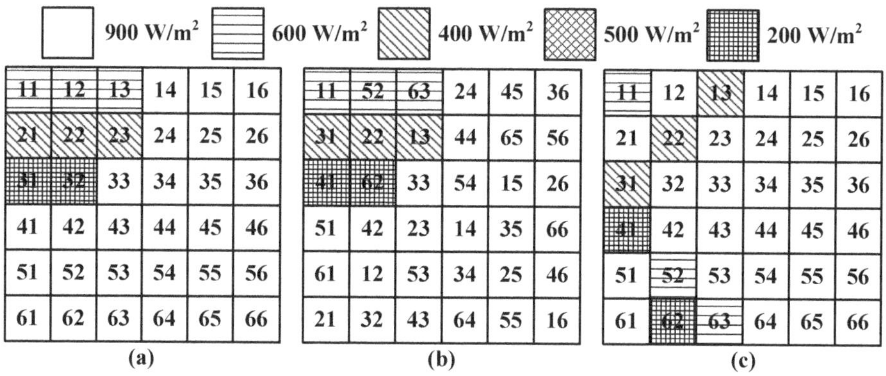

Figure 3.12 Shading pattern for Long Narrow (a) Conventional TCT configuration (b) Proposed Arrow Sudoku configuration (c) Shade dispersion with proposed configuration.

shading by distributing the shading pattern throughout the array. The comparison of the proposed configuration with different configurations gives an increment of power for the proposed configuration at 18.08% when compared to SP configuration, 11.25% when compared with BL configuration, 17.55% when compared with HC configuration, and 9.39% compared to TCT configuration.

3.3.1.4 Long Narrow Shading Conditions

This shading condition a small extension to Short Narrow shading conditions, where a smaller number of panels are shaded. Four different irradiance levels are considered under this shading condition which are $900W/m^2$, $600W/m^2$, $400W/m^2$, and $200W/m^2$. The shading pattern is shown in Fig. 3.12.

Table 3.7

Position of Maximum Power Point in Total Cross Tied and Arrow Sudoku Configuration for Long Narrow Shading Conditions

Total Cross Tied Configuration			Arrow Sudoku Configuration		
Current(A)	Voltage(V)	Power(W)	Current(A)	Voltage(V)	Power(W)
I_{R2} $4.1I_m$	$6V_m$	$24.6V_mI_m$	I_{R1} $4.6I_m$	$6V_m$	$27.6V_mI_m$
I_{R3} $4.1I_m$	$5V_m$	$20.5V_mI_m$	I_{R3} $4.6I_m$	$5V_m$	$23V_mI_m$
I_{R1} $4.5I_m$	$4V_m$	$18V_mI_m$	I_{R6} $4.6I_m$	$4V_m$	$18.4V_mI_m$
I_{R4} $5.4I_m$	$3V_m$	$16.2V_mI_m$	I_{R2} $4.9I_m$	$3V_m$	$14.7V_mI_m$
I_{R5} $5.4I_m$	$2V_m$	$10.8V_mI_m$	I_{R4} $5.1I_m$	$2V_m$	$10.2V_mI_m$
I_{R6} $5.4I_m$	V_m	$5.4V_mI_m$	I_{R5} $5.1I_m$	V_m	$5.1V_mI_m$

The currents calculations for each row of the TCT configuration are given as

$$I_{R1} = 3 \times 0.6I_m + 3 \times 0.9I_m = 4.5I_m$$
$$I_{R2} = 3 \times 0.4I_m + 3 \times 0.9I_m = 3.9I_m$$
$$I_{R3} = 2 \times 0.2I_m + 4 \times 0.9I_m = 4I_m \tag{3.14}$$
$$I_{R4} = I_{R5} = I_{R6} = 6 \times 0.9I_m = 5.4I_m$$

The currents value for each row of the proposed Arrow Sudoku configuration are given as

$$I_{R1} = 1 \times 0.6I_m + 1 \times 0.4I_m + 4 \times 0.9I_m = 4.6I_m$$
$$I_{R2} = I_{R3} = 5 \times 0.9I_m + 1 \times 0.4I_m = 4.9I_m$$
$$I_{R4} = 5 \times 0.9I_m + 1 \times 0.2I_m = 4.7I_m \tag{3.15}$$
$$I_{R5} = 5 \times 0.9I_m + 1 \times 0.6I_m = 5.1I_m$$
$$I_{R6} = 1 \times 0.2I_m + 1 \times 0.6I_m + 4 \times 0.9I_m = 4.4I_m$$

The theoretical values for Currents (under the sequence the panels are bypassed), Volt-ages and Powers are used for GMPP calculation for conventional and proposed configuration for each and every row under long wide shading conditions is tabulated in Table 3.7. The configuration helps to reshuffle the panels in such a way that the shade is dispersed across the array resulting in a power enhancement of 11% (theoretically) and 8% (through simulations) for Arrow Sudoku compared with the conventional TCT configuration. The Voltage-Current and Voltage-Power characteristics for all conventional and proposed configurations is shown in Fig. 3.13. It is easily evident from the results that the conventional configurations exhibit multiple peaks which makes it difficult to reach the maximum power at GMPP. The proposed configuration eradicates the multiple peaks thereby a single peak point is achieved.

The power is increased from 2545W to 3096W from SP to Proposed Arrow Sudoku configuration, 2837W to 3096W from TCT to Arrow Sudoku configuration, 2647W and 2648W from BL, HC configurations to proposed Arrow Sudoku configuration (3096W). The power loss occurred during shading (W & %) for SP, BL, TCT,

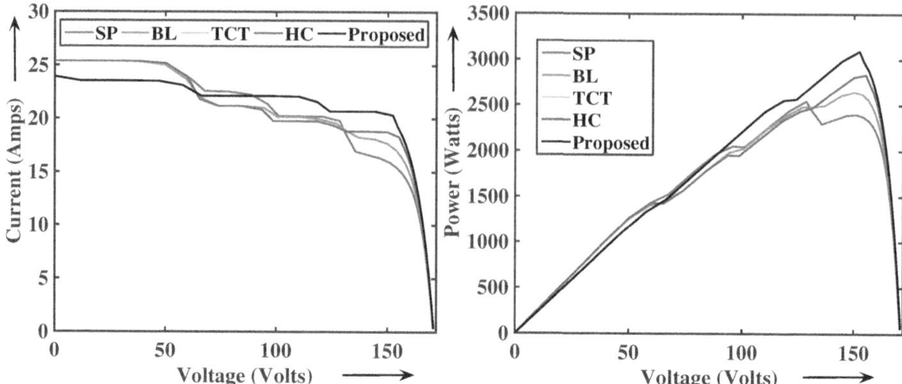

Figure 3.13 Voltage-Current and Voltage-Power characteristics for SP, BL, TCT, HC and Proposed Arrow Sudoku configurations for Long Narrow shading conditions.

Table 3.8

Generated Output Power (W) and Overall Power Loss Caused by Shading for Different Configurations under Long Narrow Shading Condition

Topology	Power (W)	Power Loss (W)	Power Loss (%)
SP	2545	1415	35.73
BL	2647	1313	33.16
TCT	2837	1123	28.36
HC	2648	1312	33.16
Proposed (AS)	**3096**	**864**	**21.82**

HC and proposed configurations are 1415W (35.73%), 1313W (33.16%), 1123W (28.36%), 1312W (33.16%) and 864W (21.82%), respectively, which are depicted in Table 3.8.

3.3.2 ANALYSIS OF THE RESULTS FOR THE PROGRESSIVE INCREMENTAL TOP TO BOTTOM MOVING SHADING CONDITIONS

The presentation of the PV Array for Progressive Incremental Top to Bottom Moving shadings along with Performance is analyzed and explained as follows

The shading pattern for Progressive Incremental Top to Bottom Moving shading conditions along with its shade dispersion is shown in Fig. 3.14. The Theoretical values for Currents, Voltages, and Powers to locate GMPP for conventional and proposed configuration for all rows under each shading case is tabulated in Table 3.9.

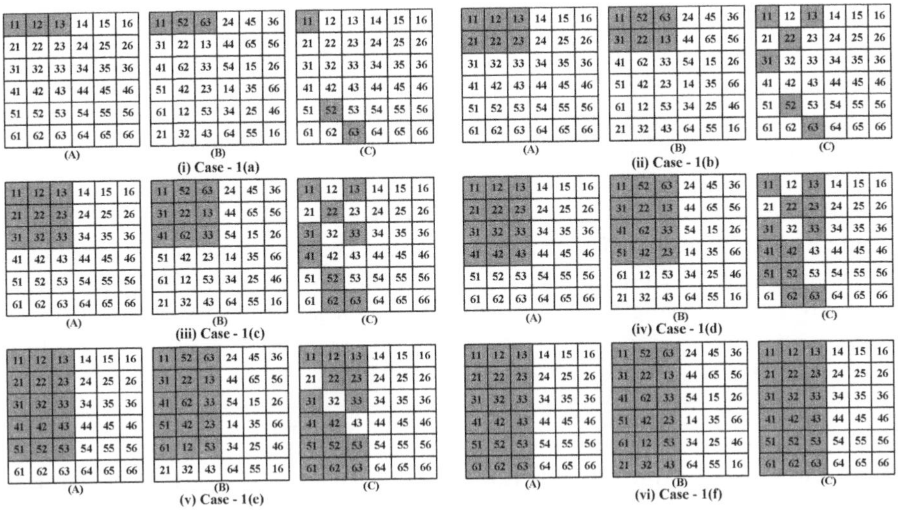

(A) (B) (C)
(i) Case - 1(a)

(A) (B) (C)
(ii) Case - 1(b)

(A) (B) (C)
(iii) Case - 1(c)

(A) (B) (C)
(iv) Case - 1(d)

(A) (B) (C)
(v) Case - 1(e)

(A) (B) (C)
(vi) Case - 1(f)

Figure 3.14 Shading pattern for 6×6 SPV Array under Progressive Incremental top to bottom Moving shading conditions (A) Shading on Conventional configuration (B) Shading on Proposed Arrow Sudoku configuration (C) Shade dispersion with proposed configuration.

3.3.2.1 Voltage-Current and Voltage-Power Characteristics with Output Power

Figure 3.15 depicts the Voltage-Current and Voltage-Power characteristic curves of SPV Array for Progressive Incremental Top to Bottom Moving shading conditions which are for the conventional and proposed configurations. The results prove the superiority of the proposed Arrow Sudoku configuration in terms of maximum output power compared with all configurations.

The Voltage-Power and Voltage-Current characteristic curves of the Solar PV Array, under conventional and proposed configurations, for case-1(a) to case-1(f), is represented in Fig. 3.14, are shown in Fig. 3.15(i–vi). The enhanced output power is obtained for the proposed Arrow Sudoku configuration against all configurations.

The conventional configurations show two peaks, one local maximum and one global maximum, for all the configurations except for Arrow Sudoku in case 1(a) of Fig. 3.15(i) in Voltage-Power and Voltage-Current characteristic curves. The proposed configuration based on Arrow Sudoku helps to better disperse the shading pattern which ultimately results in minimized mismatch of row current thus reducing the multiple peaks and the increasing the Global maximum power yielded at 3245W. The maximum output powers at global maxima are observed as 2877W for SP, 2816W for TCT, 2812W for BL, 2825W for HC, and 3245W for proposed AS configurations. These values are listed in Table 3.10 respectively. For the TCT configuration there are three shaded panels in the 1st row but the Arrow Sudoku configuration helps to displace these three panels into three different rows which helps to reduce the mismatch losses and improve the output power. Moreover, the presence of all the shaded panels in 1st row reduces the current in first row of TCT configuration but as in AS configuration current is enhanced due to dispersion of the shaded panels to the entire array thereby increasing the output power (Table 3.9).

The Voltage-Current and Voltage-Power characteristic curves for case-1(b) (Fig. 3.14(ii)) is shown in Fig. 3.15(ii). The results show that the partial shading causes disturbances in the characteristics curves, which is mitigated by Arrow Sudoku configuration, and generates highest output power. The maximum power observed at global peaks are given to be 2397W, 2463W, 2546W, 2475W, 2988W for SP, BL, TCT, HC and AS configurations respectively which are tabulated in Table 3.10.

Table 3.9

Position of Maximum Power Point in Total Cross Tied and Arrow Sudoku Configuration for Progressive Incremental Moving Shading Conditions

Total Cross Tied Configuration			Arrow Sudoku Configuration		
Current(A)	Voltage(V)	Power(W)	Current(A)	Voltage(V)	Power(W)
Case-1(a)					
I_{R1} $4.5I_m$	$6V_m$	$27V_mI_m$	I_{R1} $5.1I_m$	$6V_m$	$30.6V_mI_m$
I_{R2} $5.4I_m$	$5V_m$	$27V_mI_m$	I_{R5} $5.1I_m$	$5V_m$	$25.5V_mI_m$
I_{R3} $5.4I_m$	$4V_m$	$21.6V_mI_m$	I_{R6} $5.1I_m$	$4V_m$	$20.4V_mI_m$
I_{R4} $5.4I_m$	$3V_m$	$16.2V_mI_m$	I_{R2} $5.4I_m$	$3V_m$	$16.2V_mI_m$
I_{R5} $5.4I_m$	$2V_m$	$10.8V_mI_m$	I_{R3} $5.4I_m$	$2V_m$	$10.8V_mI_m$
I_{R6} $5.4I_m$	$1V_m$	$5.4V_mI_m$	I_{R4} $5.4I_m$	$1V_m$	$5.4V_mI_m$
Case-1(b)					
I_{R1} $4.5I_m$	$6V_m$	$27V_mI_m$	I_{R1} $4.8I_m$	$6V_m$	$28.8V_mI_m$
I_{R2} $4.5I_m$	$5V_m$	$22.5V_mI_m$	I_{R2} $5.1I_m$	$5V_m$	$25.5V_mI_m$
I_{R3} $5.4I_m$	$4V_m$	$21.6V_mI_m$	I_{R3} $5.1I_m$	$4V_m$	$20.4V_mI_m$
I_{R4} $5.4I_m$	$3V_m$	$16.2V_mI_m$	I_{R5} $5.1I_m$	$3V_m$	$15.3V_mI_m$
I_{R5} $5.4I_m$	$2V_m$	$10.8V_mI_m$	I_{R6} $5.1I_m$	$2V_m$	$10.2V_mI_m$
I_{R6} $5.4I_m$	$1V_m$	$5.4V_mI_m$	I_{R4} $5.4I_m$	$1V_m$	$5.4V_mI_m$
Case-1(c)					
I_{R1} $4.5I_m$	$6V_m$	$27V_mI_m$	I_{R1} $4.8I_m$	$6V_m$	$28.8V_mI_m$
I_{R2} $4.5I_m$	$5V_m$	$22.5V_mI_m$	I_{R3} $4.8I_m$	$5V_m$	$24V_mI_m$
I_{R3} $4.5I_m$	$4V_m$	$18V_mI_m$	I_{R6} $4.8I_m$	$4V_m$	$19.2V_mI_m$
I_{R4} $5.4I_m$	$3V_m$	$16.2V_mI_m$	I_{R2} $5.1I_m$	$3V_m$	$15.3V_mI_m$
I_{R5} $5.4I_m$	$2V_m$	$10.8V_mI_m$	I_{R4} $5.1I_m$	$2V_m$	$10.2V_mI_m$
I_{R6} $5.4I_m$	$1V_m$	$5.4V_mI_m$	I_{R5} $5.4I_m$	$1V_m$	$5.4V_mI_m$

(Continued on next page)

Table 3.9

(*Continued*)

Total Cross Tied Configuration			Arrow Sudoku Configuration		
Current(A)	**Voltage(V)**	**Power(W)**	**Current(A)**	**Voltage(V)**	**Power(W)**
Case-1(d)					
I_{R1} $4.5I_m$	$6V_m$	$27V_mI_m$	I_{R1} $4.8I_m$	$6V_m$	$28.8V_mI_m$
I_{R2} $4.5I_m$	$5V_m$	$22.5V_mI_m$	I_{R2} $4.8I_m$	$5V_m$	$24V_mI_m$
I_{R3} $4.5I_m$	$4V_m$	$18V_mI_m$	I_{R3} $4.8I_m$	$4V_m$	$19.2V_mI_m$
I_{R4} $4.5I_m$	$3V_m$	$13.5V_mI_m$	I_{R4} $4.8I_m$	$3V_m$	$14.4V_mI_m$
I_{R5} $5.4I_m$	$2V_m$	$10.8V_mI_m$	I_{R5} $4.8I_m$	$2V_m$	$9.6V_mI_m$
I_{R6} $5.4I_m$	$1V_m$	$5.4V_mI_m$	I_{R6} $4.8I_m$	$1V_m$	$4.8V_mI_m$
Case-1(e)					
I_{R1} $4.5I_m$	$6V_m$	$27V_mI_m$	I_{R1} $4.5I_m$	$6V_m$	$27V_mI_m$
I_{R2} $4.5I_m$	$5V_m$	$22.5V_mI_m$	I_{R5} $4.5I_m$	$5V_m$	$22.5V_mI_m$
I_{R3} $4.5I_m$	$4V_m$	$18V_mI_m$	I_{R6} $4.5I_m$	$4V_m$	$18V_mI_m$
I_{R4} $4.5I_m$	$3V_m$	$13.5V_mI_m$	I_{R2} $4.8I_m$	$3V_m$	$14.4V_mI_m$
I_{R5} $4.5I_m$	$2V_m$	$9V_mI_m$	I_{R3} $4.8I_m$	$2V_m$	$9.6V_mI_m$
I_{R6} $5.4I_m$	$1V_m$	$5.4V_mI_m$	I_{R4} $4.8I_m$	$1V_m$	$4.8V_mI_m$
Case-1(f)					
I_{R1} $4.5I_m$	$6V_m$	$27V_mI_m$	I_{R1} $4.5I_m$	$6V_m$	$27V_mI_m$
I_{R2} $4.5I_m$	$5V_m$	$22.5V_mI_m$	I_{R5} $4.5I_m$	$5V_m$	$22.5V_mI_m$
I_{R3} $4.5I_m$	$4V_m$	$18V_mI_m$	I_{R6} $4.5I_m$	$4V_m$	$18V_mI_m$
I_{R4} $4.5I_m$	$3V_m$	$13.5V_mI_m$	I_{R2} $4.5I_m$	$3V_m$	$13.5V_mI_m$
I_{R5} $4.5I_m$	$2V_m$	$9V_mI_m$	I_{R3} $4.5I_m$	$2V_m$	$9V_mI_m$
I_{R6} $4.5I_m$	$1V_m$	$4.5V_mI_m$	I_{R4} $4.5I_m$	$1V_m$	$4.5V_mI_m$

Figure 3.15(iii) shows the Voltage-Current and Voltage-Power characteristic curves for shading conditions given in Fig. 3.14(iii) (case-1(c)). From this Fig. 3.15 (iii), it is evident that using the conventional configurations results in multiple peaks in the characteristics curve. These multiple peaks are mitigated by employing the proposed configuration which helps to reshuffle the panels throughout the array resulting in increment of the output power. The Maximum output Powers present at global peaks are given as 2385W, 2437W, 2479W, and 2443W for SP, BL, TCT, and HC configurations, respectively, while the proposed Arrow Sudoku enhances the maximum output power to 2824W which are listed in Table 3.10. The output power shows an improved increment for the proposed Arrow Sudoku configuration because the shaded panels are dispersed throughout the array reducing the mismatch between row currents.

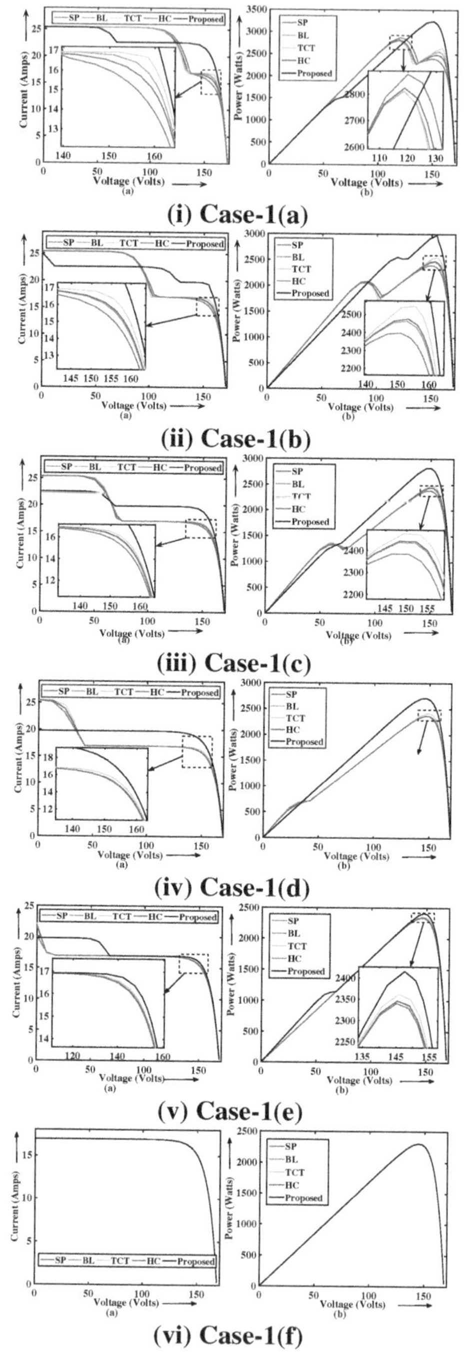

(i) Case-1(a)

(ii) Case-1(b)

(iii) Case-1(c)

(iv) Case-1(d)

(v) Case-1(e)

(vi) Case-1(f)

Figure 3.15 Voltage-Current and Voltage-Power characteristics for SP, BL, TCT, HC, and Proposed Arrow Sudoku configurations for incremental moving conditions.

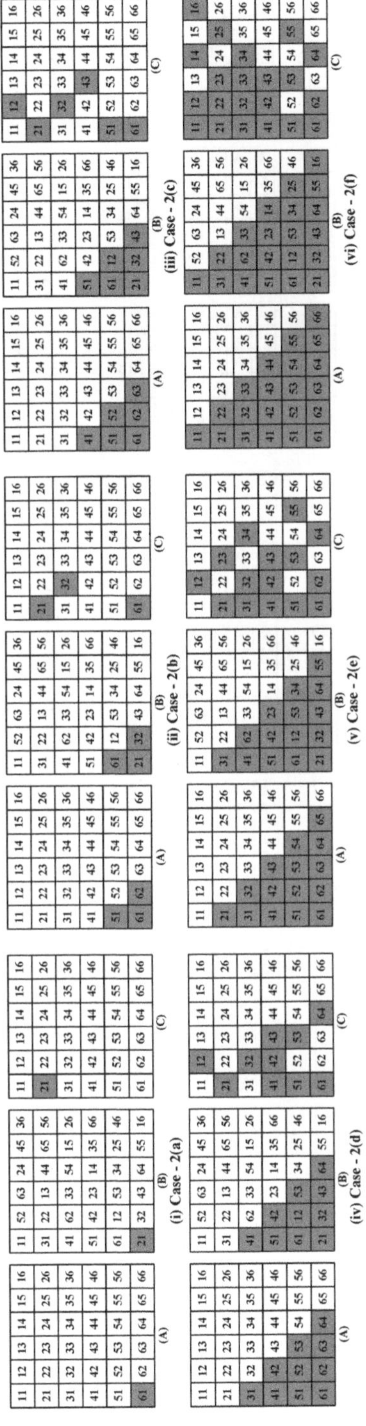

Figure 3.16 Shading pattern for 6×6 SPV Array under Progressive Incremental Diagonal Moving shading conditions (A) Shading on Conventional configuration (B) Shading on Proposed Arrow Sudoku configuration (C) Shade dispersion with proposed configuration.

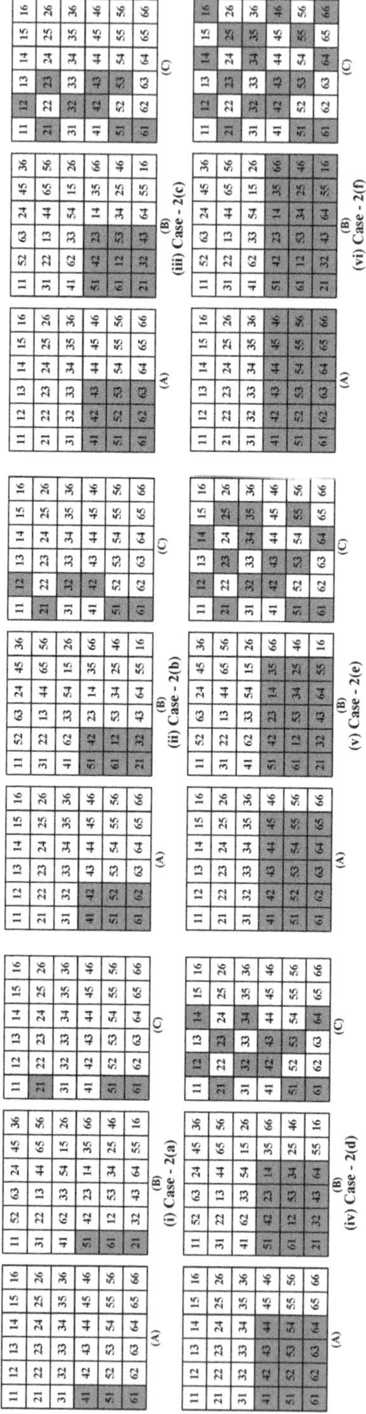

Figure 3.17 Shading pattern for 6×6 SPV Array under Progressive Incremental Left to right Moving shading conditions (A) Shading on Conventional configuration (B) Shading on Proposed Arrow Sudoku configuration (C) Shade dispersion with proposed configuration.

Table 3.10

Generated Output Power and Power Loss Caused by Shading for Conventional and Proposed Arrow Sudoku Configuration under Progressive Incremental Top to Bottom Moving Shading Conditions for 6×6 SPV Array

	Topology	Power (W)	Power Loss (W)	Power Loss (%)
Case-1(a)	SP	2877	1083	27.35
	BL	2812	1148	28.99
	TCT	2816	1144	28.89
	HC	2825	1135	28.66
	AS	**3245**	**715**	**18.06**
Case-1(b)	SP	2397	1563	39.47
	BL	2463	1497	37.8
	TCT	2546	1414	35.71
	HC	2475	1485	37.5
	AS	**2988**	**972**	**24.55**
Case-1(c)	SP	2385	1575	39.77
	BL	2437	1523	38.46
	TCT	2479	1481	37.4
	HC	2443	1517	38.31
	AS	**2824**	**1136**	**28.69**
Case-1(d)	SP	2367	1593	40.23
	BL	2370	1590	40.15
	TCT	2419	1541	38.91
	HC	2370	1590	40.15
	AS	**2713**	**1247**	**31.49**
Case-1(e)	SP	2340	1620	40.91
	BL	2346	1614	40.76
	TCT	2362	1598	40.35
	HC	2346	1614	40.76
	AS	**2416**	**1544**	**38.99**
Case-1(f)	SP	2309	1651	41.69
	BL	2309	1651	41.69
	TCT	2309	1651	41.69
	HC	2309	1651	41.69
	AS	**2309**	**1651**	**41.69**

Table 3.11

Generated Output Power and Power Loss Caused by Shading for Conventional and Proposed Arrow Sudoku Configuration under Progressive Incremental Diagonal Moving Shading Conditions for 6×6 SPV Array

	Maximum Output Power (W)					Power Losses (W)				
	SP	BL	TCT	HC	Proposed (AS)	SP	BL	TCT	HC	Proposed (AS)
Case-2(a)	3161	3279	3429	3290	**3429**	378	260	110	249	**110**
Case-2(b)	2780	2882	3030	2892	**3246**	759	658	509	648	**293**
Case-2(c)	2394	2479	2614	2502	**2989**	1145	1060	925	1037	**551**
Case-2(d)	2008	2086	2171	2101	**2784**	1531	1453	1369	1438	**755**
Case-2(e)	1654	1688	1742	1682	**2416**	1886	1852	1797	1858	**1123**
Case-2(f)	1303	1371	1380	1367	**1998**	2236	2168	2160	2173	**1541**

Table 3.12

Generated Output Power and Power Loss Caused by Shading for Conventional and Proposed Arrow Sudoku Configuration under Progressive Incremental Left to Right Moving Shading Conditions for 6×6 SPV Array

	Maximum Output Power (W)					Power Losses (W)				
	SP	BL	TCT	HC	Proposed (AS)	SP	BL	TCT	HC	Proposed (AS)
Case-3(a)	3158	3204	3246	3189	**3414**	381	335	293	350	**125**
Case-3(b)	2773	2797	2877	2776	**3127**	766	742	663	763	**412**
Case-3(c)	2385	2437	2479	2443	**2824**	1154	1102	1060	1096	**715**
Case-3(d)	2005	2033	2074	2044	**2713**	1534	1506	1465	1495	**826**
Case-3(e)	1621	1655	1657	1628	**2416**	1918	1884	1882	1911	**1123**
Case-3(f)	1240	1240	1240	1240	**2309**	2299	2299	2299	2299	**1230**

The Voltage-Power and Voltage-Current characteristic curves for cases-1(d) and 1(e) (Fig. 3.14(iv) and Fig. 3.14(v)) are shown in Fig. 3.15(iv) and Fig. 3.15(v), respectively. It is highlighted that the distortions as a result of the partial shading in conventional configurations are minimized by employing the proposed configuration which ultimately leads to improvement in output power. The maximum output power obtained for SP, BL, TCT, HC, and proposed AS are observed as 2367W, 2370W, 2419W, 2370W, and 2713W for case 1(d) and 2340W, 2346W, 2362W, 2346W, and 2416W for case 1(e), respectively. The output power improvement for the Proposed Arrow Sudoku is obtained because of physical reshuffling of the shaded and unshaded panels. For the final case-1(f), all the configurations generate maximum output power of 2309W because shading pattern covers half of the array as shown in Fig. 3.14(vi). The Voltage-Current and Voltage-Power characteristic curves are given in Fig. 3.15(vi). The maximum output power is tabulated in Table 3.10.

3.3.2.2 Mismatch Power Loss from Normal Conditions to the Shading Conditions

The proposed configuration for Progressive Incremental Top to Bottom Moving shading conditions yields very less mismatch power loss (Table 3.10) against the other configurations. A maximum mismatch power loss of 10% is obtained for case-1(c) where under this condition 25% panels are covered in shading pattern.

The maximum power obtained and its mismatch power loss for progressive incremental diagonal moving shading conditions (Fig. 3.16) and left to right (Fig. 3.17) shading conditions are given in Table 3.11 and Table 3.12, respectively. From the tables, it can be seen that the proposed method is giving better results than all the existing methods for all the moving and static shading conditions.

3.4 SUMMARY

This chapter proposes a work based on Arrow Sudoku puzzle pattern configuration scheme for SPV array to minimize the effect of partial shading conditions. The proposed configuration follows static configuration scheme under which the panels are reshuffled physically across the array without disturbing the electrical connections. The relocation of the panels helps to reduce the mismatch between row currents by achieving uniform dispersion of concentrated shading pattern on to the entire array. To validate the proposed reconfiguration, it was compared with the conventional configurations (SP, BL, TCT, and HC). The results were evaluated for static and moving shading conditions. The Static Shading conditions include Short Wide, Short Narrow, Long Wide, and Long Narrow shading conditions and different moving shading condition which included the continuous increment in shadows. The usefulness of the proposed method is well established through the results obtained under all shading conditions. Thus, the results confirm that the proposed configuration is better against existing configurations because it generates enhanced output power along with reducing the mismatch between row currents. This method also fails to scale for any size of SPV Array, i.e., it works only for 6×6 SPV Array. In order to overcome this disadvantage, a new algorithm is proposed to enhance the output parameters for all the sizes of SPV Array.

4 Array Reconfiguration with Muddled Baker Map

4.1 INTRODUCTION

The methods suggested in chapter 2 and chapter 3 were incapable of attaining inclusive shade diffusion for the arrangement as the panels were scattered among similar rows and columns and could not be scalable to any size of an arrangement. Hence, there is a requirement for competent repositioning methods that can capably work under any size of an arrangement and decrease the output power losses (POL) under shading scenarios. To take on these needs, a new image preprocessing-based configuration method is suggested in this effort. A widely held image encryption method, called Muddled Baker Map (MBM) is utilized which repositions the pixels of an image. Every panel in the SPV (solar photovoltaic) arrangement is treated as being identical to a single image pixel. Under this method, the electrical contacts of each panel present in the Solar-Photovoltaic arrangement remain intact but physically the position of the panel in the TCT configuration (C_{TCT}) is altered permanently. The positions of the panels are changed within various rows and columns under the suggested reconfiguration fashion to improve the performance parameters for the partially shaded panels. The effectiveness of the suggested MBM reconfiguration method is validated using 4 × 4, 6 × 6, and 8 × 8 Solar-Photovoltaic arrangements for various non-uniform and progressive incremental shading scenarios. The results are evaluated against previously Standard known interconnection methods which are Series-Parallel (C_{SP}), Bridge-Linked (C_{BL}), Total-Cross-Tied (C_{TCT}), and Honey-Comb (C_{HC}) configurations. The results confirmed that the power produced at the output (PO) is significantly enhanced with the suggested image processing-based MBM method under various non-uniform and progressive incremental shading scenarios.

4.1.1 NOVELTY OF THE PROPOSED APPROACH

The suggested configuration method is advanced to the traditional configurations because of the following:

1. Contrary to the majority of the currently used reconfiguration methods, the suggested reconfiguration adjusts the panels along different rows and columns, which substantially reduces the concentrated shade.
2. The suggested MBM arrangement reduces the disparity in the level of row currents by a substantially higher margin when compared to the previous approaches because of the shade pattern's improved dispersion.
3. This suggested work is best suited for various shading scenarios, owing to its random reconfigurability.

DOI: 10.1201/9781003285830-4

The rest part of the chapter is structured as follows: Section 4.2 describes the methodology of the suggested approach for arrangement reconfiguration. Section 4.3 discusses about the comprehensive results of the all the configurations used for the study. Finally, the chapter is concise in Section 4.4

4.2 METHODOLOGY OF THE PROPOSED APPROACH

Under image processing, an image is divided into the form of a matrix comprising pixels as its elements [100]. To enhance security, image encryption is executed by altering the positions of pixels in the image [101]. The position of the pixel is altered based on the algorithm key which leads to only end users decoding it. The most popular method for encrypting a picture in image processing is MBM. The fundamental benefit of using MBMs over other Muddled maps is that they make it simple to build optimum algorithm keys in relation to the wide range of potential keys [102–104]. Under this approach, the map displays an action on the unit square (N x N), and the formula below [105] defines the mathematical representation of Baker Map 'B'.

$$B(a,b) = \begin{cases} (a/2, 2b, & 0 \le b < 1/2 \\ (a/2 + 1/2, 2b, & 1/2 \le b \le 1 \end{cases} \tag{4.1}$$

As illustrated in Fig. 4.1(a), the map represents action on the unit square. In order to form a rectangle, the vertical column on the left $[0, 1/2) \times [0, 1)$ is stressed horizontally and contracted vertically. A similar mapping is performed on the vertical column on the right from $[0, 1) \times [0, 1/2)$ to $[0, 1) \times [1/2, 1)$ as presented in Fig. 4.1(a). The Baker Map is a disorganized bijection of $N \times N$ (the unit square) up on itself.

As a substitute for the separation into two similar rectangles, it is separated based on MBM. The Baker Map is characterized by $B(v_1, v_2, ..., v_m)$, where the vector $[v_1, v_2, ..., v_m]$ denotes the algorithm key. The conditions that must be fulfilled by the fundamentals of an algorithm key are described below:

- The algorithm key has 'm' elements that divide an image's total number of pixels into 'm' rectangles.
- The fundamentals in the algorithm key should satisfy $v_1 + v_2 + ... + v_k = N$, where N is the number of pixels in each row.
- The fundamentals preset in the algorithm key $[v_1, v_2, ..., v_k]$ should be the multiples of N i.e., N should be precisely separable by V_i, where $1 \le i \le k$.

The selection of algorithm key is made so that each integer V_i divides 'N' and $V_i = v_1 + v_2 + ... + v_i$ with $V_i \le l < V_i + v_i$ where, the new position of the pixel (l', s') is given by ([106])

$$B(v_1, v_2, ..., v_m)(l', s') = \left(\frac{N}{v_i}(l - V_i) + s \mod \left(\frac{N}{v_i} \right), \frac{v_i}{N} \left(s - s \mod \left(\frac{N}{v_i} \right) \right) + V_i \right) \tag{4.2}$$

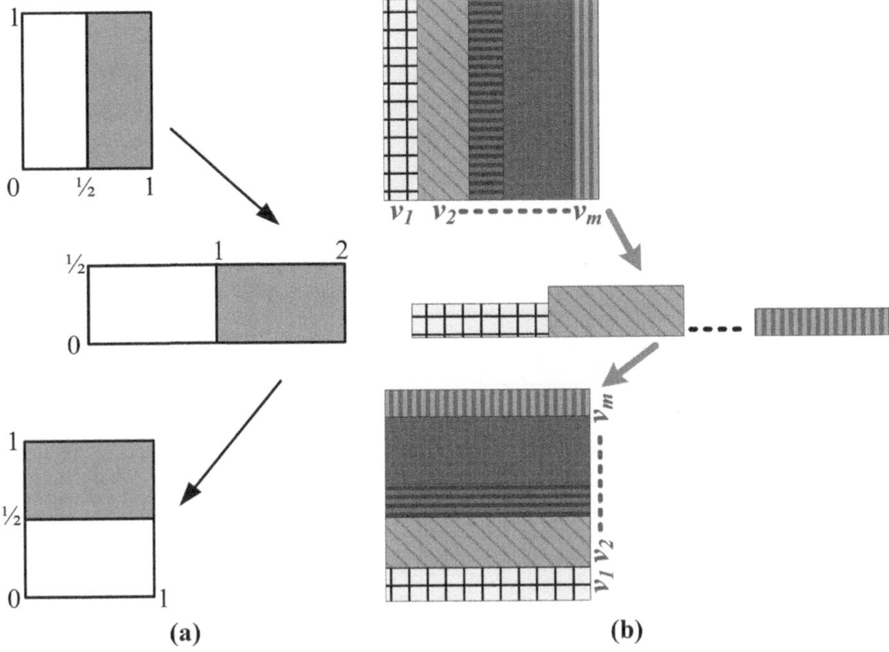

Figure 4.1   Muddled Baker Map: (a) Two-Dimensional Baker Map (b) Muddled Baker Map.

Under the comprehensive Baker Map, every rectangle extends horizontally with a factor of $1/v_i$. Simultaneously, the rectangle is reduced vertically by the aspect of v_i. In the end, Bernoulli's shifts cause each rectangle to be stacked on top of the others as presented in Fig. 4.1(b)

The generalized Baker Map is depicted in Fig. 4.1(b), and the phases that result in Baker Map randomization are as follows:

1. The matrix is separated into 'k' primary rectangles and every rectangle takes 'N' elements with width 'v_i' where, $N = v_1 + v_2 + \ldots + v_k$.
2. From right to left, study the rectangles from top to bottom.
3. Each rectangle is unwrapped to a row matrix by scanning its elements bottom to top from left to right.
4. The row matrices developed are positioned in row-wise method to acquire the final reconfigured arrangement.

The process of the dividing $N \times N$ SPV array with 'k' rectangles is clearly explained with a flowchart as shown in Fig. 4.2(a). The progression of Muddled Baker Map reconfiguration matrix of $N \times N$ SPV array is presented in Fig. 4.2(b).

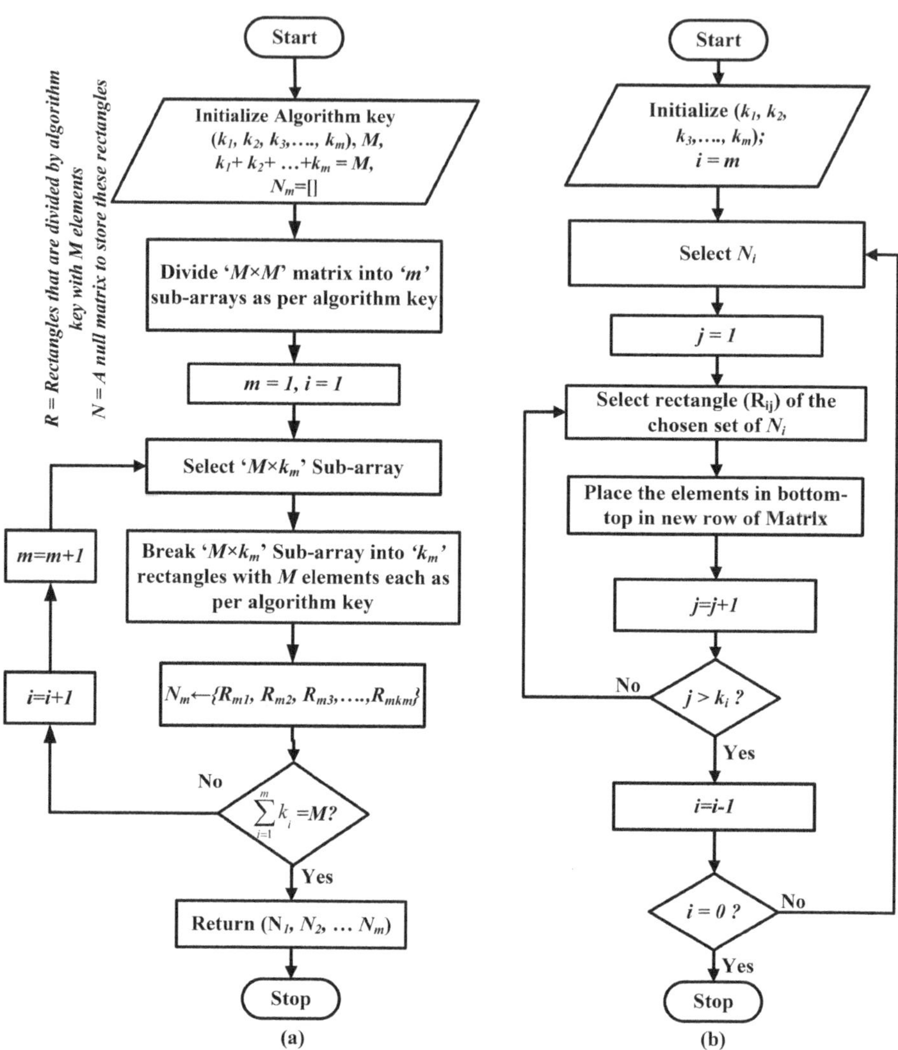

Figure 4.2 Flowchart for process flow of Muddled Baker Map (a) Flowchart for division of an $N \times N$ SPV array into 'k' rectangles with an algorithm key $[v_1, v_2, ..., v_k]$, (b) Flowchart for evolution of Muddled Baker Map reconfiguration of $N \times N$ SPV array.

4.2.1 RECONFIGURATION OF 4×4 SPV ARRAY THROUGH MUDDLED BAKER MAP

For 4×4 SPV array, the algorithm key is $(1, 2, 1)$, which means, $N = 4$, $v_1 = 1$, $v_2 = 2$ and $v_3 = 1$. The reconfiguration method for 4×4 Solar-Photovoltaic arrangement is displayed in Fig. 4.3. In the first stage, the 4×4 SPV array is separated into three

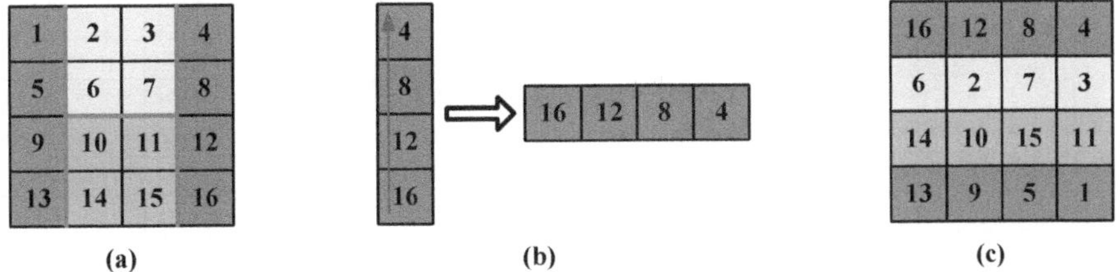

Figure 4.3 Reconfiguration through Muddled Baker Map for a 4×4 SPV array (a) Dividing the matrix into rectangles using algorithm key $(1,2,1)$, (b) Sample unwrapping of elements in sub-blocks to rows, and (c) Final reconfigured array.

main rectangles each with one, two and one columns singly based on algorithm key. Hence, primary rectangles are formed with array sizes, 4×1, 4×2, and 4×1, respectively. Subsequently, for the second phase, the three main rectangles developed would be further subdivided into several sub-blocks each having four elements, as number of rows and columns considered under this work are four. The first rectangle is then divided into multiple sub-blocks with four elements each. The second rectangle can be separated into 4×1 or 2×2 sub-blocks, but 4×1 doesn't meet the main requirement of algorithm key (splitting the columns based on algorithm key). So, 2×2 is the only sub-block which fulfils all the situations. So, this main rectangle of 4×2 is separated into two sub-blocks each having two columns and two rows (of arrangement size 2×2) each. It can be clearly seen that the first and third rectangle is split so that the algorithm key state is assured as well as sub-block covers four elements. Figure 4.3(a) clearly depicts the separation of arrangement into rectangles and with algorithm key. Later, all of the second stage sub-blocks are unpacked to produce successive row matrices that meet the constraints listed below. Starting from the right, the sub-blocks are examined from top to bottom. As seen in Fig. 4.3(b) for 4×4 SPV array each sub-block is now transformed into a row matrix by scanning its elements from bottom to top and from left to right. The same steps are followed for the rest of the sub-blocks and the resulting row matrices are assembled together to develop the randomized arrangement of size 4×4 as shown in Fig. 4.3(c).

4.2.2 RECONFIGURATION OF 6×6 SPV ARRAY THROUGH MUDDLED BAKER MAP

The algorithm key for a 6×6 Solar-Photovoltaic arrangement is $(3,1,2)$, which translates to $N = 6$, $v_1 = 3$, $v_2 = 1$, and $v_3 = 2$. Figure 4.4 shows the 6×6 Solar-Photovoltaic arrangement reconfiguration method. On the basis of the algorithm key, the 6×6 Solar-Photovoltaic arrangement is divided into three basic rectangles, each with three, one, and two columns. As a result, primary rectangles are created for the arrangement sizes of 6×3, 6×1, and 6×2, respectively. The three main rectangles were then further divided into several sub-blocks, each of which had six elements, at the second step. This basic rectangle, which measures 6×3, is broken up into

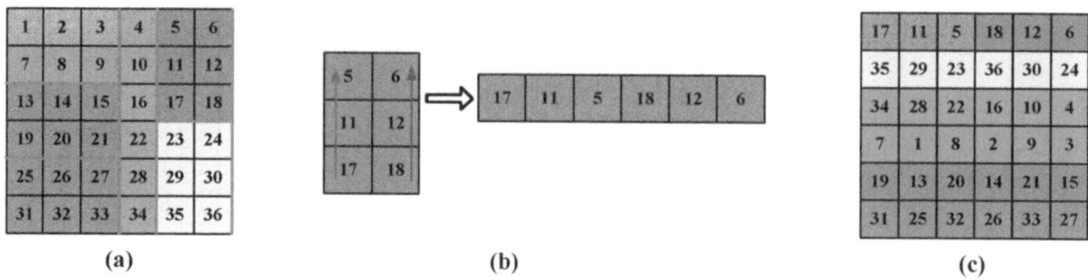

Figure 4.4 Reconfiguration through Muddled Baker Map for a 6×6 SPV array (a) Dividing the matrix into rectangles using algorithm key $(3, 1, 2)$, (b) Sample unwrapping of elements in sub-blocks to rows, and (c) Final reconfigured array.

three smaller rectangles, each of which has three columns and two rows (in an arrangement of size 2×3). Similarly, the second and third rectangles are split into sub-blocks having one column and six rows (sub block of size 6×1) and into two sub-blocks consisting of two number of columns and three number of rows each (of arrangement size 3×2) respectively. Figure 4.4(a) clearly depicts the separation of arrangement into rectangles along with the algorithm key. In the 2nd step, these sub-blocks are unpacked to develop row matrices as shown in Fig. 4.4(b) for 6×6 Solar-Photovoltaic arrangement. The same steps is observed for the rest of the sub-blocks and the resultant row matrices are packed together to develop the randomized arrangement of size 6×6 as shown in Fig. 4.4(c).

4.2.3 RECONFIGURATION OF 8×8 SPV ARRAY THROUGH MUDDLED BAKER MAP

For 8×8 SPV array, the algorithm key is $(2, 4, 2)$ which relates to $N = 6$, $v_1 = 2$, $v_2 = 4$, and $v_3 = 2$. For the first step, the 8×8 array is divided into three primary rectangles of array sizes, '8×2', '8×4' and '8×2', respectively. Subsequently, at the second stage, each rectangle is partitioned into small rectangles such that it contains 8 elements i.e., the first primary rectangle is partitioned into two sub-blocks consisting of two number of columns and four number of rows (of array size '4×2') each and subsequently second rectangle is split for four sub-blocks having four number of columns and two number of rows each (each sub block of size '2×4') and third primary rectangle into two sub-blocks comprising of two number of columns and four number of rows each (of array size '4×2') which are depicted in gray and yellow colors, respectively. Figure 4.5(a) clearly depicts the division of array into rectangles and with algorithm key. In the second stage, unwrapping of sub-blocks is performed as shown in Fig. 4.5(b). The resultant row matrices are packed together to form a row in the randomized array of size 8×8 as shown in Fig. 4.5(c).

Figures 4.6, 4.7, and 4.8 show the structural arrangement of conventional TCT configuration and proposed Muddled Baker Map reconfiguration with new position of the panels in 4×4, 6×6, and 8×8 arrays, respectively. It is observed from these figures that only the physical position of panels is transformed and the electrical

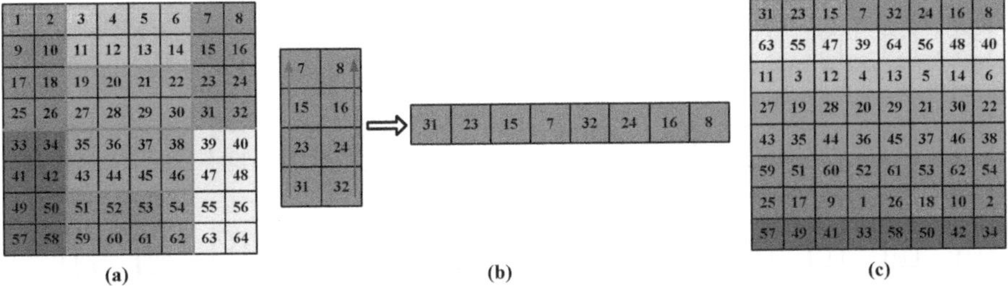

Figure 4.5 Reconfiguration through Muddled Baker Map for a 8×8 SPV array (a) Dividing the matrix into rectangles using algorithm key $(2,4,2)$, (b) Sample unwrapping of elements in sub-blocks to rows, and (c) Final reconfigured array.

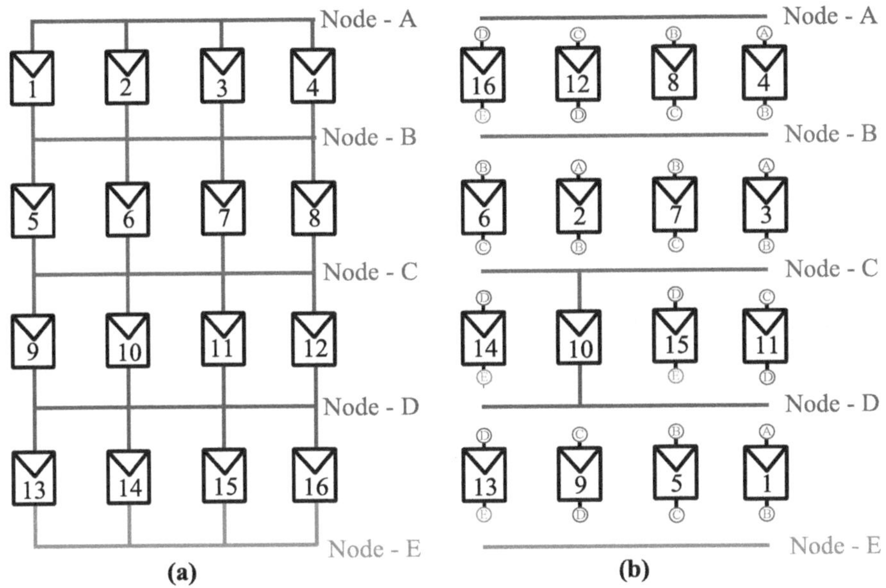

Figure 4.6 Structural arrangement: (a) Conventional TCT configuration and (b) Proposed Muddled Baker Map reconfiguration for 4×4 SPV array.

circuitry is same as that of C_{TCT}. The nodes are represented on each side of the panel, which shows that they are connected electrically to the even row. The panel number '5' is in second row and first column in the standard C_{TCT} (Fig. 4.6(a)) of 4×4 Solar-Photovoltaic arrangement, which is repositioned physically based on suggested MBM to fourth row and third column without changing the connection electrically, i.e., in between Node-B and Node-C which is clearly observed in Fig. 4.6(b). In the standard TCT setup, the panel number '4' for 6×6 Solar-Photovoltaic arrangement is in the first row and fourth column ((Fig. 4.7(a))) which is repositioned physically

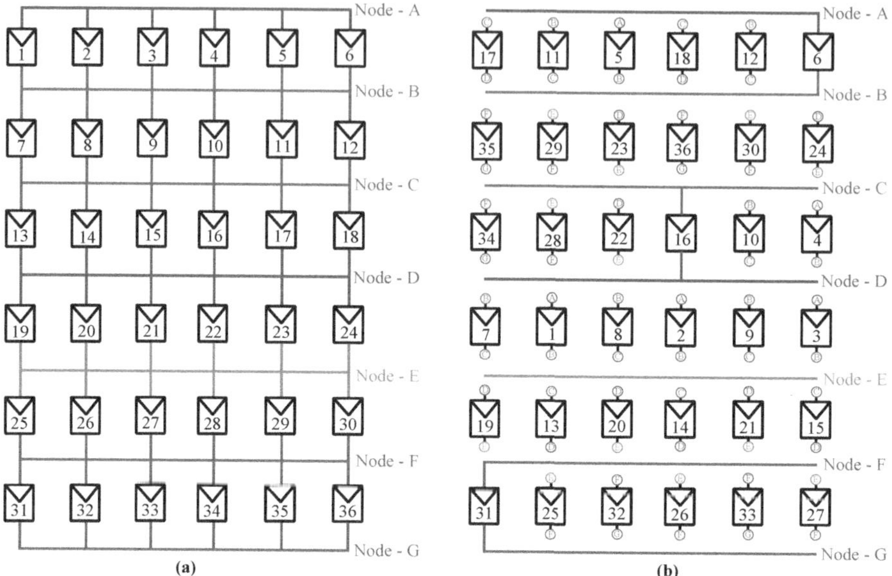

Figure 4.7 Structural arrangement: (a) Conventional TCT configuration and (b) Proposed Muddled Baker Map reconfiguration for 6×6 SPV array.

based on the suggested MBM to third row and sixth column without modifying the electric assembly i.e., in between Node-A and Node-B which is evidently seen in Fig. 4.7(b). Similar to this, in an 8×8 Solar-Photovoltaic arrangement, panel number '24' which was formerly in the third row and eighth column (Fig. 4.8(a)), is now in the first row and third column, but it still has the same electrical connections as a C_{TCT} between Nodes C and D, as illustrated in Fig. 4.8(b).

4.3   RESULTS AND DISCUSSIONS

The suggested configuration is validated for various non-varying and incremental shading scenarios for 4×4, 6×6, and 8×8 SPV arrays. The contrast was based upon produced P_O and P_{OL} compared with the Standard (which includes the C_{SP}, C_{BL}, C_{HC}, and C_{TCT}) configurations.

4.3.1   RESULTS UNDER NON-UNIFORM SHADING CONDITIONS

As mentioned in Section **??**, four static shading conditions (Short Narrow, Short Wide, Long Wide, and Long Narrow) are used in this work to test the efficacy of the proposed work for 4×4, 6×6, and 8×8 is shown in Figs. 4.9, 4.10, and 4.11, respectively.

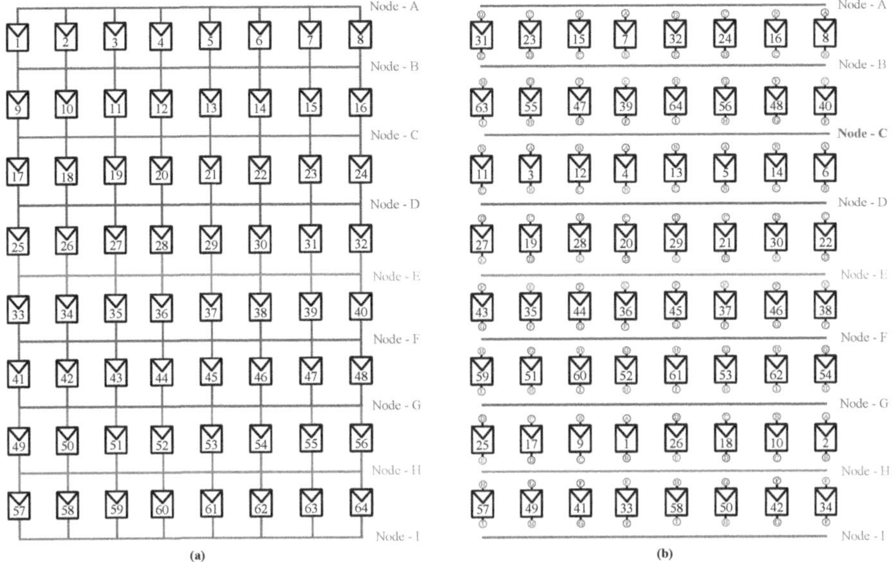

Figure 4.8 Structural arrangement: (a) Conventional TCT configuration and (b) Proposed Muddled Baker Map reconfiguration for 8 × 8 SPV array.

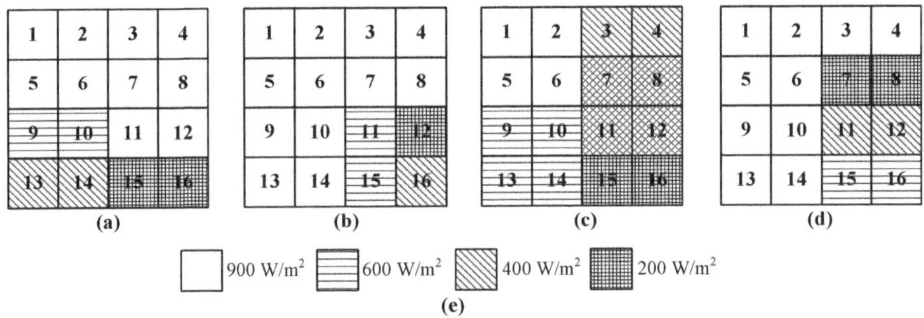

Figure 4.9 Shading profile for 4 × 4 under Non-Uniform shading conditions used in this work (a) Short Wide (b) Short Narrow (c) Long Wide (d) Long Narrow (e) Irradiation levels.

4.3.1.1 For 4 × 4 **SPV Array Under Non-Uniform Shading Conditions**

Under non-uniform shading scenarios, the suggested MBM design is examined for a 4 × 4 Solar-Photovoltaic arrangement and compared to the standard configurations.

4.3.1.1.1 Short Wide Shading Scenarios

The Solar-Photovoltaic arrangement's panels are exposed to four different groups of irradiation intensities under SW shading scenarios. According to Fig. 4.12,

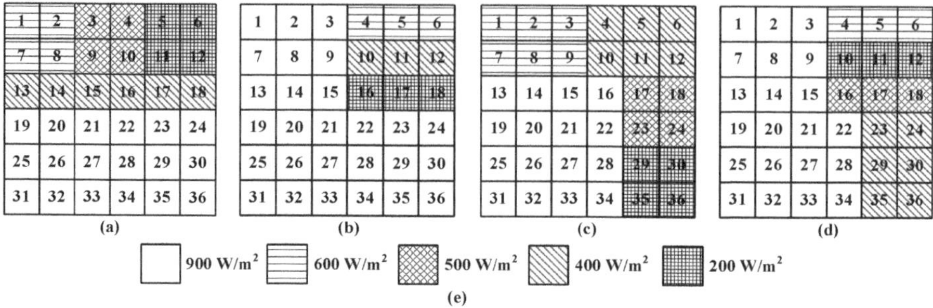

Figure 4.10 Shading profile for 6×6 under Non-Uniform shading conditions (a) Short Wide (b) Short Narrow (c) Long Wide (d) Long Narrow (e) irradiation levels.

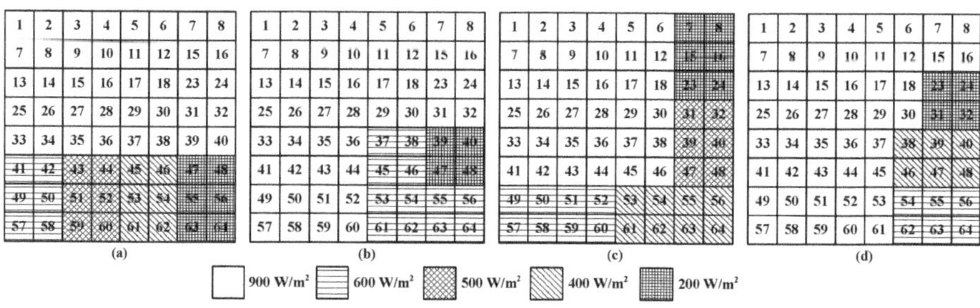

Figure 4.11 Shading profile for 8×8 under (a) Short Wide (b) Short Narrow (c) Long Wide (d) Long Narrow Non-Uniform shading conditions.

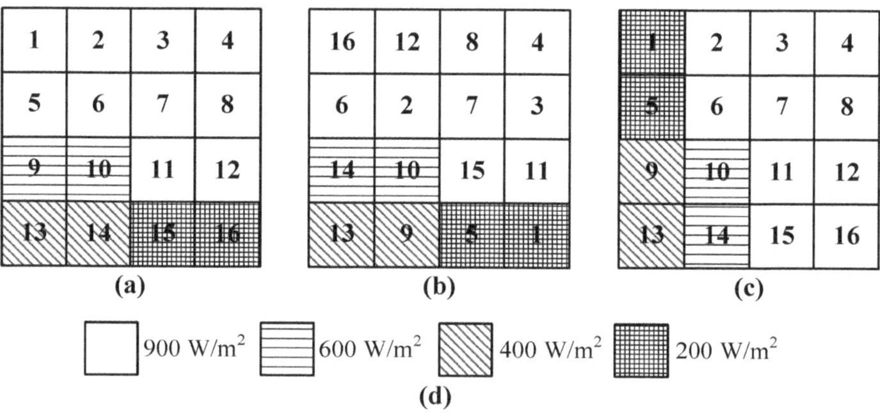

Figure 4.12 Short wide shading conditions for 4×4 SPV array: (a) TCT configuration (b) Proposed Muddled Baker Map configuration (c) Shade dispersion with the proposed configuration (d) Different level of irradiations.

Table 4.1

Position of Maximum Power Point in Total Cross Tied and Muddled Baker Map Configuration for Short Wide Shading Conditions

Total Cross Tied Configuration			Muddled Baker Map Configuration		
Current(A)	Voltage(V)	Power(W)	Current(A)	Voltage(V)	Power(W)
I_{R4} $1.2I_m$	$4V_m$	$4.8V_mI_m$	I_{R4} $2.8I_m$	$4V_m$	$11.2V_mI_m$
I_{R3} $3I_m$	$3V_m$	$9V_mI_m$	I_{R3} $2.8I_m$	$3V_m$	$8.4V_mI_m$
I_{R2} $3.6I_m$	$2V_m$	$7.2V_mI_m$	I_{R2} $2.9I_m$	$2V_m$	$5.8V_mI_m$
I_{R1} $3.6I_m$	V_m	$3.6V_mI_m$	I_{R1} $2.9I_m$	V_m	$2.9V_mI_m$

Group-1 received radiation at a level similar to Standard test settings ($900\text{W}/\text{m}^2$), while Groups 2, 3, and 4 received $600\text{W}/\text{m}^2$, $400\text{W}/\text{m}^2$, and $200\text{W}/\text{m}^2$, accordingly.

Calculating each row current of the Solar-Photovoltaic arrangement is necessary in order to theoretically create the position of the Global Maximum Power Point (GMPP) on characteristic curves. The panels in rows 1 and 2 of the TCT arrangement receive the same amount of $900\text{W}/\text{m}^2$ of radiation under these shading scenarios. These rows' current output can be represented as follows:

$$I_{R1} = I_{R2} = 4 \times 0.9I_m = 3.6I_m \tag{4.3}$$

In row-3, two panels are exposed to $600\text{W}/\text{m}^2$ and the other two with $900\text{W}/\text{m}^2$ irradiation. The current through is given by

$$I_{R3} = (2 \times 0.6)I_m + (2 \times 0.9)I_m = 3I_m \tag{4.4}$$

Similarly, in row-4, two panels are irradiated with $200\text{W}/\text{m}^2$ and the further two with $400\text{W}/\text{m}^2$ irradiation. The current at the output is developed on this row is given by

$$I_{R4} = (2 \times 0.2)I_m + (2 \times 0.4)I_m = 1.2I_m \tag{4.5}$$

From eqs. (4.3) to (4.5), it is detected that the row currents are changing in the range of $1.2I_m$ to $3.6I_m$. As a result of the large changes in row currents, several peaks are produced. Table 4.1 lists the various electrical characteristics as well as the position of the highest power point. The highest power generated under no-shading scenarios is provided as ($900\text{W}/\text{m}^2$)

$$P_a = V_aI_a = (4 \times 0.9)I_m \times 4V_m = 14.4V_mI_m, \tag{4.6}$$

where V_m represents the maximum voltage and $4V_m$ represents the voltage at the output of the Solar-Photovoltaic arrangement, which is due to four panels attached in series.

For C_{TCT}, the overall current at the output developed at the Solar-Photovoltaic arrangement will only be $1.2I_m$, for these shading scenarios, because of series connections present in the rows. Due to the bypass diode not being used, the output voltage is $4V_m$.

Under these shading circumstances, the P_O generated by the TCT setup, without bypassing the panels, is evaluated as,

$$P_{a,TCT} = (1.2)I_m \times 4V_m = 4.8V_m I_m \tag{4.7}$$

For the suggested MBM configuration, as depicted in Fig. 4.12(c), the concentrated shade on last two rows is spread across the whole Solar-Photovoltaic arrangement due to which it reduces the consequence of SW shading. For the suggested MBM, the panels in rows 1 and 2 and rows 3 and 4 receive the same irradiation levels under these shading scenarios after dispersing the shade. The current developed for the rows can be calculated as:

$$I_{R1} = I_{R2} = (1 \times 0.2)I_m + (3 \times 0.9)I_m = 2.9I_m \tag{4.8}$$

$$I_{R3} = I_{R4} = (1 \times 0.4)I_m + (1 \times 0.6)I_m + (2 \times 0.9)I_m = 2.8I_m \tag{4.9}$$

The eqs. (4.8) and (4.9), show that the row currents developed are of value $2.8I_m$ and $2.9I_m$. The suggested MBM configuration helps to minimize the row currents mismatch which tends to reduce the multiple peaks. The electrical parameters and the GMPP for MBM method shown in Table 4.1

Therefore, the output (P_O) developed by proposed MBM configuration is specified as

$$P_{a,MBM} = (2.8)I_m \times 4V_m = 11.2V_m I_m \tag{4.10}$$

Based on the eqs. (4.10) and (4.7), it can be compared that the P_O developed by suggested MBM gives GMPP of $11.2V_m I_m$ when contrasted with C_{TCT} ($4.8V_m I_m$). Figure 4.13 shows the characteristic curves for Voltage against Current and Voltage against Power for all the configurations stated above. It is clear that reducing the mismatch current in the rows leads to a reduction in multiple peaks in the suggested configuration and improves the current and power values at the highest power point. The suggested MBM method generates a maximum PO of 1224W when compared against SP (894W), BL (917W), TCT (947W), and HC (911W). It is examined from the outcomes that a power improvement of 22.63% is achieved when the configuration is changed from TCT to MBM.

4.3.1.1.2 Short Narrow Shading Scenarios

A smaller number of panels are shadowed for the SN shading scenarios. Four diverse stages of irradiation that includes $900W/m^2$, $600W/m^2$, $400W/m^2$, and $200W/m^2$ are applied on the Solar-Photovoltaic arrangement presented in Fig. 4.14.

The value of currents developed in each row under C_{TCT} is calculated as

$$I_{R1} = I_{R2} = 4 \times 0.9I_m = 3.6I_m \tag{4.11}$$

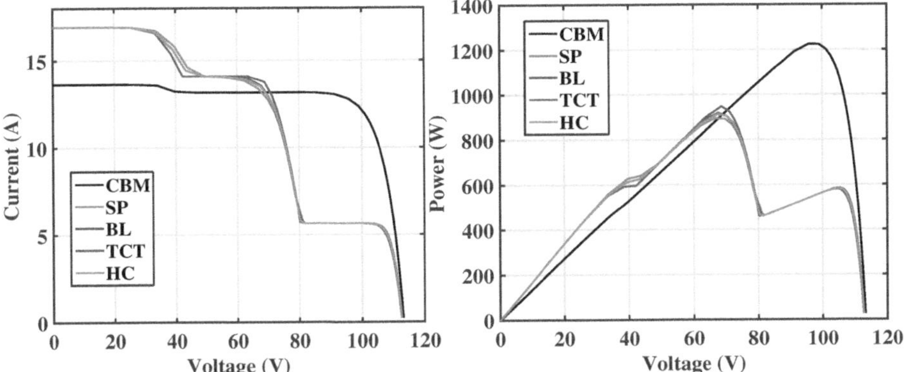

Figure 4.13 Voltage-Current and Voltage-Power Characteristic curves for 4×4 SPV array under short wide shading conditions.

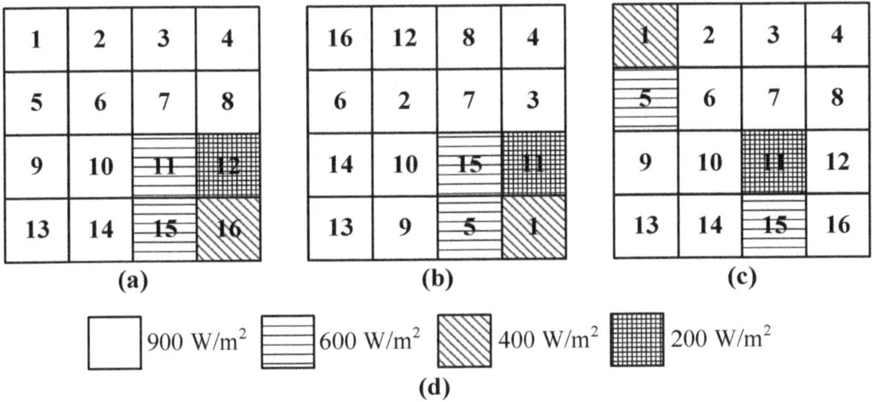

Figure 4.14 Short Narrow shading conditions for 4×4 SPV array: (a) TCT configuration (b) Proposed Muddled Baker Map configuration (c) Shade dispersion with the proposed configuration (d) Different level of irradiations.

$$I_{R3} = (2 \times 0.9)I_m + (1 \times 0.6)I_m + (1 \times 0.2)I_m = 2.6I_m \tag{4.12}$$

$$I_{R4} = (2 \times 0.9)I_m + (1 \times 0.6)I_m + (1 \times 0.4)I_m = 2.8I_m \tag{4.13}$$

For the above stated shading scenario, the power developed at the output node under C_{TCT} are assessed as

$$P_{a,TCT} = 2.6I_m \times 4V_m = 10.4V_m I_m \tag{4.14}$$

The value of current developed in each row under the suggested MBM configuration is estimated as

$$I_{R1} = (3 \times 0.9)I_m + (1 \times 0.4)I_m = 3.1I_m \tag{4.15}$$

Table 4.2

Position of Maximum Power Point in Total Cross Tied and Muddled Baker Map Configuration for Short Narrow Shading Conditions

Total Cross Tied Configuration			Muddled Baker Map Configuration		
Current(A)	**Voltage(V)**	**Power(W)**	**Current(A)**	**Voltage(V)**	**Power(W)**
I_{R3} $2.6I_m$	$4V_m$	$10.4V_mI_m$	I_{R3} $2.9I_m$	$4V_m$	$11.6V_mI_m$
I_{R4} $2.8I_m$	$3V_m$	$8.4V_mI_m$	I_{R1} $3.1I_m$	$3V_m$	$9.3V_mI_m$
I_{R2} $3.6I_m$	$2V_m$	$7.2V_mI_m$	I_{R2} $3.3I_m$	$2V_m$	$6.6V_mI_m$
I_{R1} $3.6I_m$	V_m	$3.6V_mI_m$	I_{R4} $3.3I_m$	V_m	$3.3V_mI_m$

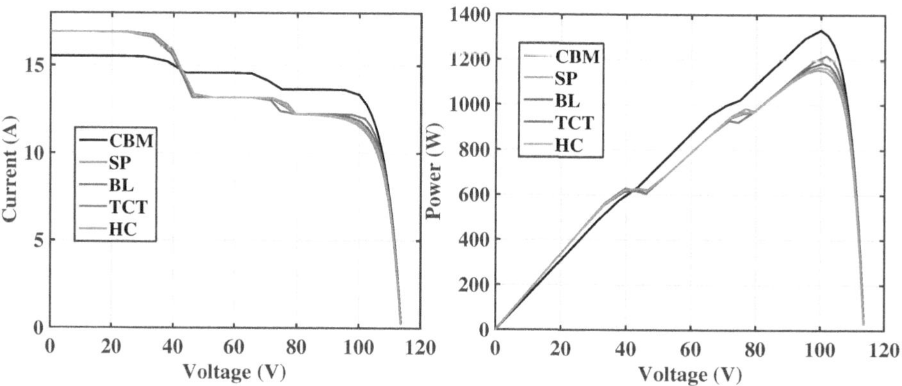

Figure 4.15 Voltage-Current and Voltage-Power Characteristic curves for 4×4 SPV array under short narrow shading conditions.

$$I_{R2} = I_{R4} = (3 \times 0.9)I_m + (1 \times 0.6)I_m = 3.3I_m \tag{4.16}$$

$$I_{R3} = (3 \times 0.9)I_m + (1 \times 0.2)I_m = 2.9I_m \tag{4.17}$$

For the above stated shading scenario, the power developed at the output node under suggested configuration are assessed as

$$P_{a,MBM} = 2.9I_m \times 4V_m = 11.6V_mI_m \tag{4.18}$$

There is a substantial improvement in the power at the output under the suggested configuration ($10.4V_mI_m$) against the standard C_{TCT} ($11.6V_mI_m$) which is shown in eqs. (4.14) and (4.18). The theoretical location of the maximum PO is tabulated in Table 4.2.

Figure 4.15 displays the characteristic curves for the Solar-Photovoltaic arrangement under short narrow shading scenarios.

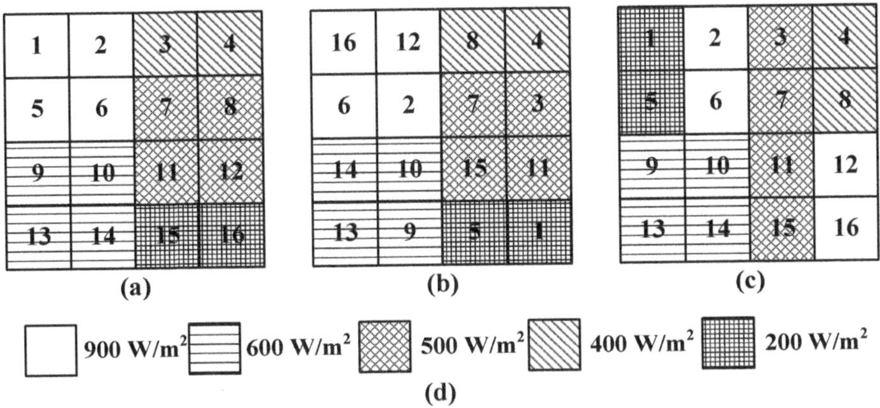

Figure 4.16 Long Wide shading conditions for 4×4 SPV array: (a) TCT configuration (b) Proposed Muddled Baker Map configuration (c) Shade dispersion with the proposed configuration (d) Different level of irradiations.

The least power is developed by SP (1155W), followed by HC (1166W) and BL (1186W). The P_O for C_{TCT}, a robust benchmark configuration, is 1219W. When compared to the C_{TCT}, the suggested configuration produces a maximum PO of 1333W, which results in an 8.5% increase in power.

4.3.1.1.3 Long Wide Shading Conditions

The three-fourth of the panels are covered with five diverse irradiation levels from $900W/m^2$ to $200W/m^2$ for this shading scenario as shown in Fig. 4.16. The maximum power level for a 4×4 Solar-Photovoltaic arrangement against the suggested MBM configuration and Standard TCT is shown in Table 4.3 for LW Shading scenarios.

The row currents under C_{TCT} for each row is assessed as

$$I_{R1} = (2 \times 0.9)I_m + (2 \times 0.4)I_m = 2.6I_m \tag{4.19}$$

$$I_{R2} = (2 \times 0.9)I_m + (2 \times 0.5)I_m = 2.8I_m \tag{4.20}$$

$$I_{R3} = (2 \times 0.6)I_m + (2 \times 0.5)I_m = 2.2I_m \tag{4.21}$$

$$I_{R4} = (2 \times 0.6)I_m + (2 \times 0.2)I_m = 1.6I_m \tag{4.22}$$

Under this shading scenario, the power developed at the output for C_{TCT} is assessed as

$$P_{a,TCT} = 1.6I_m \times 4V_m = 6.4V_mI_m \tag{4.23}$$

The value of current developed in each row under the suggested MBM configuration is estimated as

$$I_{R1} = I_{R2} = (1 \times 0.2)I_m + (1 \times 0.9)I_m + (1 \times 0.5)I_m + (1 \times 0.4)I_m = 2I_m \tag{4.24}$$

Table 4.3

Position of Maximum Power Point in Total Cross Tied and Muddled Baker Map Configuration for Long Narrow Shading Conditions

Total Cross Tied Configuration			Muddled Baker Map Configuration		
Current(A)	Voltage(V)	Power(W)	Current(A)	Voltage(V)	Power(W)
I_{R4} $1.6I_m$	$4V_m$	$6.4V_mI_m$	I_{R1} $2I_m$	$4V_m$	$8V_mI_m$
I_{R3} $2.2I_m$	$3V_m$	$6.6V_mI_m$	I_{R2} $2I_m$	$3V_m$	$6V_mI_m$
I_{R1} $2.6I_m$	$2V_m$	$5.2V_mI_m$	I_{R3} $2.6I_m$	$2V_m$	$5.2V_mI_m$
I_{R2} $2.8I_m$	V_m	$2.8V_mI_m$	I_{R4} $2.6I_m$	V_m	$2.6V_mI_m$

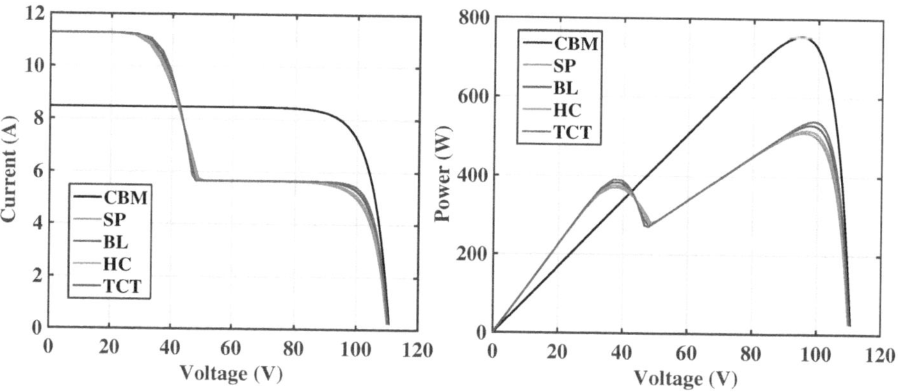

Figure 4.17 Voltage-Current and Voltage-Power Characteristic curves for 4×4 SPV array under long wide shading conditions.

$$I_{R3} = I_{R4} = (2 \times 0.6)I_m + (1 \times 0.5)I_m + (1 \times 0.9)I_m = 2.6I_m \qquad (4.25)$$

The power generated at the output node under the suggested configuration for the above stated shading scenarios are evaluated as

$$P_{a,MBM} = 2I_m \times 4V_m = 8V_mI_m \qquad (4.26)$$

The characteristic curve which includes the Voltage against Current and Voltage against Power plots under LW shading scenarios for the Solar-Photovoltaic arrangement are presented in Fig. 4.17. The suggested MBM method gives maximum PO of 756W against SP (511W), BL (531W), TCT (541W), and HC (517W). Based on the results, it can be concluded that switching from TCT to MBM resulted in a power improvement of 28.4%.

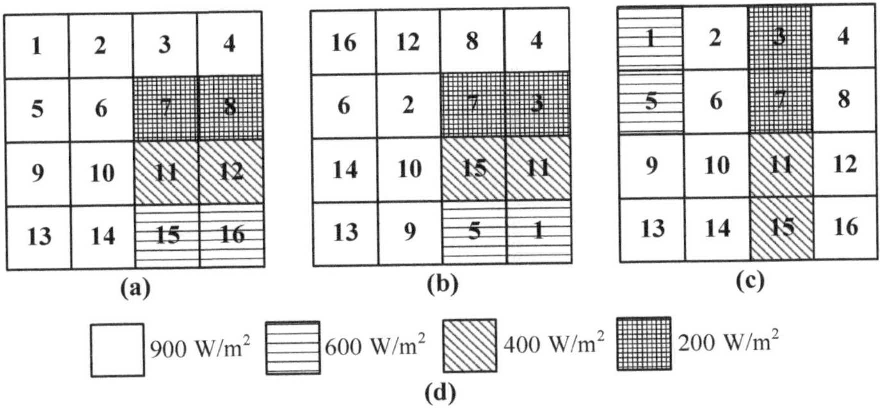

Figure 4.18   Long Narrow shading conditions for 4×4 SPV array: (a) TCT configuration (b) Proposed Muddled Baker Map configuration (c) Shade dispersion with the proposed configuration (d) Different level of irradiations.

4.3.1.1.4   Long Narrow Shading Scenarios

Compared to the SN shading scenarios, many more panels are shaded under the LN shading scenarios. Four diverse irradiation levels of $900\mathrm{W/m^2}$, $600\mathrm{W/m^2}$, $400\mathrm{W/m^2}$ and $200\mathrm{W/m^2}$ are applied on the SPV array as presented in Fig. 4.18.

The Voltage against Current and Voltage against Power characteristic curves for the Solar-Photovoltaic arrangement under above stated shading scenario are depicted in Fig. 4.19. The C_{SP} produces the lowest P_O of 972W followed by BL which generates P_O of 994W which is very closely trailed by HC with P_O of 995W. The standard C_{TCT} outputs P_O of 1037W. The suggested configuration generates a maximum P_O of 1171W which causes the power boost of 11.44% when MBM configuration is evaluated against C_{TCT}. The calculated GMPP at the output is tabularized in Table 4.4.

The power reduction for a 4×4 Solar-Photovoltaic arrangement with non-uniform shading scenarios is shown in Fig. 4.20 for a variety of configurations, including C_{SP}, C_{BL}, C_{TCT}, C_{HC}, and the suggested MBM. The graph depicts that loss in power for the MBM is lowest compared against the known standard configurations. The loss in power in C_{SP} is highest trailed by C_{BL}, C_{HC}, and C_{TCT} configurations while the suggested MBM configuration registers the lowest mismatch losses.

4.3.1.2   For 6×6 **SPV Array Under Non-Uniform Shading Conditions**

For non-uniform shading scenarios, the suggested MBM configuration is tested on a 6×6 Solar-Photovoltaic arrangement and contrasted with the standard configurations. To exhibit the effectiveness of the suggested approach, same non-uniform shading pattern is employed as used for the 4×4 Solar-Photovoltaic arrangement. The non-uniform shading scenarios used for 6×6 Solar-Photovoltaic arrangement is shown in Fig. 4.10.

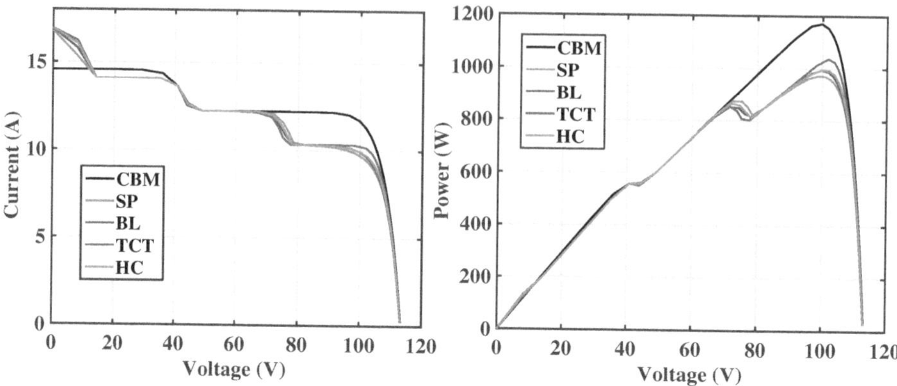

Figure 4.19 Voltage-Current and Voltage-Power Characteristic curves for 4×4 SPV array under long narrow shading conditions.

Table 4.4

Position of Maximum Power Point in Total Cross Tied and Muddled Baker Map Configuration for Long Narrow Shading Conditions

Total Cross Tied Configuration			Muddled Baker Map Configuration		
Current(A)	**Voltage(V)**	**Power(W)**	**Current(A)**	**Voltage(V)**	**Power(W)**
I_{R2} $\quad 2.2I_m$	$4V_m$	$8.8V_mI_m$	I_{R1} $\quad 2.6I_m$	$4V_m$	$10.4V_mI_m$
I_{R3} $\quad 2.6I_m$	$3V_m$	$7.8V_mI_m$	I_{R2} $\quad 2.6I_m$	$3V_m$	$7.8V_mI_m$
I_{R4} $\quad 3I_m$	$2V_m$	$6V_mI_m$	I_{R3} $\quad 3.1I_m$	$2V_m$	$6.2V_mI_m$
I_{R1} $\quad 3.6I_m$	V_m	$3.6V_mI_m$	I_{R4} $\quad 3.1I_m$	V_m	$3.1V_mI_m$

The irradiation levels vary in terms of diverse shapes and sizes based on the type of shading under various non-uniform shading scenarios, such as SW, SN, LW, and LN. The outcomes for all these configurations are presented in Table 4.5 and it is concluded that power developed under the suggested MBM configuration is highest when assessed against C_{SP}, C_{BL}, C_{TCT}, and C_{HC} with minimal P_{ML}. A power boost of 33%, 19%, 10%, and 13% is obtained for the suggested configurations against standard CTCT under SW, SN, LW, and LN shading scenarios for 6×6 Solar-Photovoltaic arrangement.

4.3.1.3 For 8×8 SPV Array Under Non-Uniform Shading Conditions

Under these non-uniform shading scenarios, the suggested MBM arrangement is also examined and contrasted with Standard configurations on an 8×8 Solar-Photovoltaic

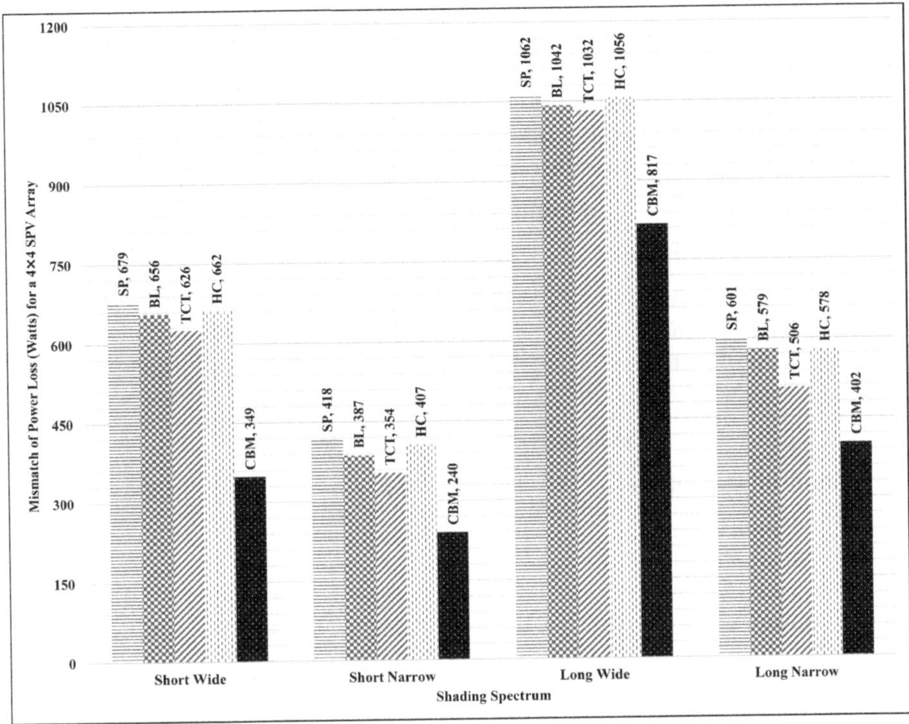

Figure 4.20 Power loss (in Watts) with 4×4 SPV array for conventional and proposed Muddled Baker Map configurations under different Non-Uniform shading conditions.

arrangement. The non-uniform shading scenarios used for testing of 8×8 Solar-Photovoltaic arrangement are presented in Fig. 4.11. The outcomes are presented in Table 4.6 for different configurations, and when contrasted with C_{SP}, C_{BL}, C_{TCT}, and C_{HC}, it has been found that the power generated by the suggested MBM configuration is producing the highest energy yield. In contrast to the standard C_{TCT}, 21% power improvement is obtained in SW with a power loss of 37.91%, 12% power increase is observed in Short Narrow shading conditions with a power loss of 19.63%, 24.4% power raise is obtained in Long Wide shading conditions with a power loss of 45.5% and 14% power improvement is obtained in Long Narrow shading conditions obtaining a power loss of 27.15%.

4.3.2 PROGRESSIVE INCREMENTAL MOVING SHADING CONDITIONS

The 4×4, 6×6, and 8×8 SPV arrays are subjected to different incremental shadings that are moving from one direction to another. The results are explained as follows:

Table 4.5

Output Power and Mismatch Power Loss by 6×6 SPV Array for Different Conventional and Proposed Muddled Baker Map Configurations under Various Non-Uniform Shading Conditions

Output Power (in Watts)					
Shading Conditions	**SP**	**BL**	**HC**	**TCT**	**MBM**
Short Wide	1470	1508	1490	1707	**2525**
Short Narrow	2308	2286	2324	2387	**2958**
Long Wide	1828	2086	2027	2095	**2332**
Long Narrow	2104	2310	2293	2363	**2713**

Mismatch Power Loss (in Watts)					
Shading Conditions	**SP**	**BL**	**HC**	**TCT**	**MBM**
Short Wide	2069	2031	2049	1832	**1014**
Short Narrow	1231	1253	1215	1152	**581**
Long Wide	1711	1453	1512	1444	**1207**
Long Narrow	1435	1229	1246	1176	**826**

Mismatch Power Loss (in Percentage)					
Shading Conditions	**SP**	**BL**	**HC**	**TCT**	**MBM**
Short Wide	58.46	57.39	57.9	51.77	**28.65**
Short Narrow	34.78	35.41	34.33	32.55	**16.42**
Long Wide	48.35	41.06	42.72	40.8	**34.11**
Long Narrow	40.55	34.73	35.21	33.23	**23.34**

4.3.2.1 For 4×4 SPV Array Under Progressive Incremental Top to Bottom Moving Shading Conditions

The Solar-Photovoltaic arrangement efficacy over progressive incremental Top to Bottom moving shading scenarios is analyzed using performance metrics in terms of maximum P_O generated and P_{ML} and are explained as follows:

4.3.2.1.1 Voltage-Power and Voltage-Current Characteristics and Maximum Output Power

The Voltage against Power and Voltage against Current characteristics of the arrangement, for standard and suggested configurations, for different cases (case-1(a) to case-1(d)) under progressive incremental Top to Bottom moving shading scenarios, as portrayed in Fig. 4.21, are shown in Fig. 4.22. Table 4.7 illustrates the electric factors for all cases and the results in terms of power generated and mismatch loss obtained with various configurations under these scenarios are shown in Table 4.8.

Table 4.6

Output Power and Mismatch Power Loss by 8×8 SPV Array for Different Conventional and Proposed Muddled Baker Map Configurations under Various Non-Uniform Shading Conditions

Output Power (in Watts)					
Shading Conditions	**SP**	**BL**	**HC**	**TCT**	**MBM**
Short Wide	3219	3223	3229	3220	**3907**
Short Narrow	4076	4620	4251	4521	**5057**
Long Wide	2537	2623	2978	2757	**3430**
Long Narrow	3980	4015	4206	4012	**4584**
Mismatch Power Loss (in Watts)					
Shading Conditions	**SP**	**BL**	**HC**	**TCT**	**MBM**
Short Wide	3073	3069	3063	3072	**2385**
Short Narrow	2216	1672	2041	1771	**1235**
Long Wide	3755	3669	3314	3535	**2862**
Long Narrow	2312	2277	2086	2280	**1708**
Mismatch Power Loss (in Percentage)					
Shading Conditions	**SP**	**BL**	**HC**	**TCT**	**MBM**
Short Wide	48.84	48.78	48.68	48.82	**37.91**
Short Narrow	35.22	26.57	32.44	28.15	**19.63**
Long Wide	59.68	58.31	52.67	56.18	**45.49**
Long Narrow	36.75	36.19	33.15	36.24	**27.15**

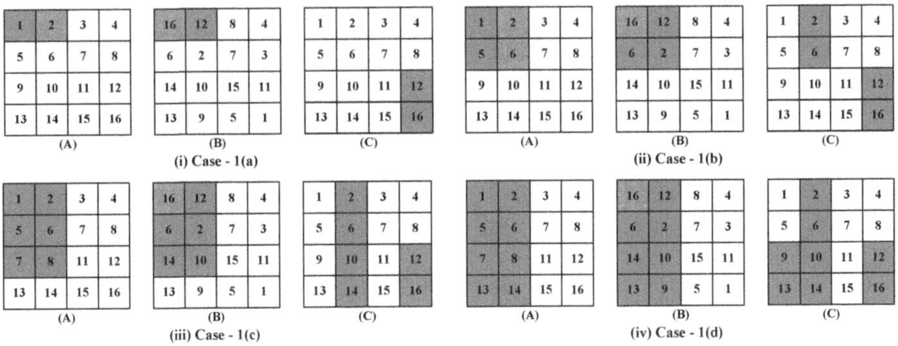

Figure 4.21 Shading pattern for 4×4 SPV Array under progressive incremental Top to Bottom moving shading conditions (A) Shading on conventional configuration (B) Shading on proposed Muddled Baker Map configuration (C) Shade dispersion with proposed configuration.

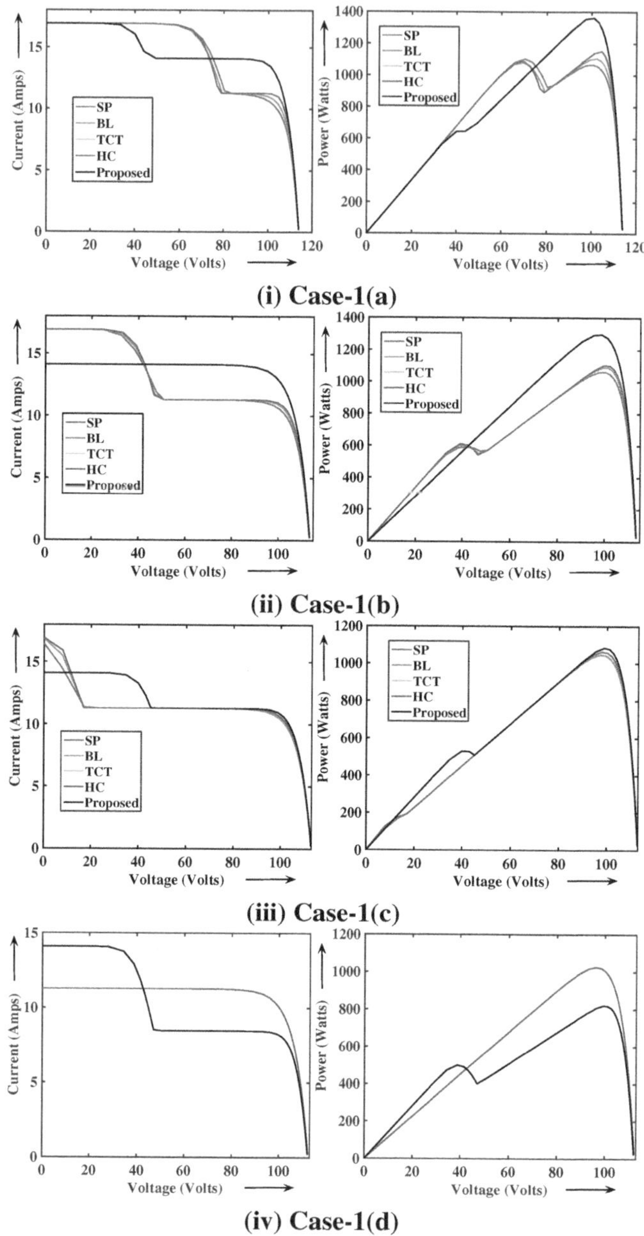

Figure 4.22 Voltage-Current and Voltage-Power characteristics for SP, BL, TCT, HC, and proposed Muddled Baker Map configurations under incremental top to bottom moving conditions for 4×4 SPV array.

Table 4.7

Theoretical Calculation of Position of Maximum Power Point in Conventional and Proposed Muddled Baker Map Configurations under Progressive Incremental Top to Bottom Moving Shading Conditions for 4×4 SPV Array

Total Cross Tied Configuration			Muddled Baker Map Configuration		
Current(A)	Voltage(V)	Power(W)	Current(A)	Voltage(V)	Power(W)
Case - 1(a)					
I_{R1} $2.4I_m$	$4V_m$	$9.6V_mI_m$	I_{R1} $3.6I_m$	$4V_m$	$14.4V_mI_m$
I_{R2} $3.6I_m$	$3V_m$	$10.8V_mI_m$	I_{R2} $3.6I_m$	$3V_m$	$10.8V_mI_m$
I_{R3} $3.6I_m$	$2V_m$	$7.2V_mI_m$	I_{R3} $3I_m$	$2V_m$	$6V_mI_m$
I_{R4} $3.6I_m$	V_m	$3.6V_mI_m$	I_{R4} $3I_m$	V_m	$3V_mI_m$
Case - 1(b)					
I_{R1} $2.4I_m$	$4V_m$	$9.6V_mI_m$	I_{R1} $3I_m$	$4V_m$	$12V_mI_m$
I_{R2} $2.4I_m$	$3V_m$	$7.2V_mI_m$	I_{R2} $3I_m$	$3V_m$	$9V_mI_m$
I_{R3} $3.6I_m$	$2V_m$	$7.2V_mI_m$	I_{R3} $3I_m$	$2V_m$	$6V_mI_m$
I_{R4} $3.6I_m$	V_m	$3.6V_mI_m$	I_{R4} $3I_m$	V_m	$3V_mI_m$
Case - 1(c)					
I_{R1} $2.4I_m$	$4V_m$	$9.6V_mI_m$	I_{R1} $2.4I_m$	$4V_m$	$9.6V_mI_m$
I_{R2} $2.4I_m$	$3V_m$	$7.2V_mI_m$	I_{R2} $2.4I_m$	$3V_m$	$7.2V_mI_m$
I_{R3} $2.4I_m$	$2V_m$	$4.8V_mI_m$	I_{R3} $3I_m$	$2V_m$	$6V_mI_m$
I_{R4} $3.6I_m$	V_m	$3.6V_mI_m$	I_{R4} $3I_m$	V_m	$3V_mI_m$
Case - 1(d)					
I_{R1} $2.4I_m$	$4V_m$	$9.6V_mI_m$	I_{R1} $1.8I_m$	$4V_m$	$7.2V_mI_m$
I_{R2} $2.4I_m$	$3V_m$	$7.2V_mI_m$	I_{R2} $1.8I_m$	$3V_m$	$5.4V_mI_m$
I_{R3} $2.4I_m$	$2V_m$	$4.8V_mI_m$	I_{R3} $3I_m$	$2V_m$	$6V_mI_m$
I_{R4} $2.4I_m$	V_m	$2.4V_mI_m$	I_{R4} $3I_m$	V_m	$3V_mI_m$

For case-1(a), the characteristic curves illustrate that the suggested MBM shadow dispersion outperforms more standard configurations in terms of PO as depicted in Fig. 4.22(i). Under these shading scenarios, the P_O is 1103W for C_{SP}, 1104W for C_{BL}, 1130W for C_{HC} and the commonly used C_{TCT} produced an PO of 1151W while the suggested MBM reconfiguration makes an improved P_O of 1361W. Another strong reflection from Fig. 4.22(i) is that multiple peaks and steps are generated in Voltage against Current and Voltage against Power characteristic curves respectively. With the shade dispersion through MBM, the numerous peaks are reduced, and the GMPP yielded only once.

Table 4.8

Generated Output Power and Power Loss Caused by Shading for Conventional and Proposed Muddled Baker Map Configuration under Progressive Incremental Top to Bottom Moving Shading Conditions for 4×4 SPV Array

	Topology	Power (W)	Power Loss (W)	Power Loss (%)
Case – 1(a)	SP	1103	470	29.88
	BL	1104	469	29.82
	TCT	1151	422	26.83
	HC	1130	443	28.16
	Proposed (MBM)	**1361**	**212**	**13.48**
Case – 1(b)	SP	1060	513	32.61
	BL	1087	486	30.9
	TCT	1101	472	30.01
	HC	1068	505	32.1
	Proposed (MBM)	**1297**	**276**	**17.55**
Case – 1(c)	SP	1046	527	33.5
	BL	1044	529	33.63
	TCT	1062	511	32.49
	HC	1053	520	33.06
	Proposed (MBM)	**1084**	**489**	**31.09**
Case – 1(d)	SP	1026	547	34.77
	BL	1026	547	34.77
	TCT	1026	547	34.77
	HC	1026	547	34.77
	Proposed (MBM)	**820**	**753**	**47.87**

The Voltage-Current and Voltage-Power characteristic curves for case-1(b) are depicted in Fig. 4.22(ii). The rising level of power (P_O) developed for different configurations is C_{SP} with 1060W, C_{HC} with 1068W, C_{BL} with 1087W, and C_{TCT} with 1101W and the proposed MBM-b based reconfiguration is obtained at 1297W and these are tabulated in Table 4.8. The four panels presented for shaded rows under Standard configurations are divided into two rows and two columns but in MBM configuration, these four shaded panels are dissolved into four diverse rows which helps to disperse the shade. Moreover, the current in first row of TCT is $2.4I_m$ but in MBM configuration, current is improved to $3.6I_m$ thereby increasing the power (Table 4.7).

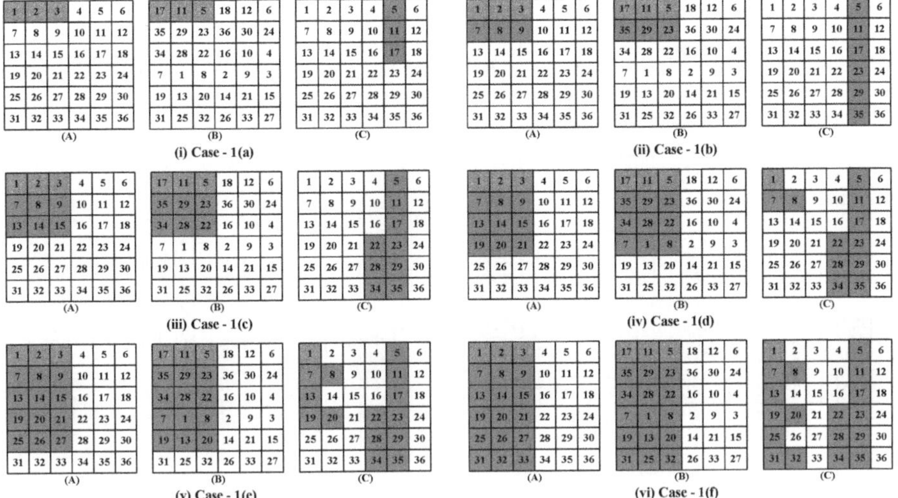

Figure 4.23 Shading pattern for 6×6 SPV array under progressive incremental Top to Bottom moving shading conditions (A) Shading on conventional configuration (B) Shading on proposed Muddled Baker Map configuration (C) Shade dispersion with proposed configuration.

As above-mentioned, under this case there is a gradual increment in terms of rows than against previous case-1(b). The shading scenario with shade dispersion is presented in Fig. 4.21(iii). The theoretical PO is same in both the cases under these shading scenarios as tabularized in Table 4.7. The various configurations generate the P_O of 1046W with C_{SP}, 1044W with C_{BL}, 1053W with C_{HC}, 1062W with C_{TCT} and 1084W with the suggested MBM configuration. The Voltage against Current and Voltage against Power characteristic curves for this shading are presented in Fig. 4.22(iii).

In the last case-1(d), all the left-side panels in an array are shaded and the right-side panels are unshaded. The shading scenario with shade dispersion is shown in Fig. 4.21(iv). For this case, all the Standard configurations develop equal level of maximum P_O of 1026W, and is closely followed by the suggested configuration of 820W. The highest power level at the output along with the mismatch losses are tabularized in Table 4.8.

4.3.2.1.2 *Mismatch Power Loss from Normal Conditions to the Shading Scenarios*

When contrasted with other configurations for the same shading scenario, the suggested configuration for progressive incremental Top to Bottom moving shading circumstances gives lessen P_{ML} (Table 4.8). The highest P_{ML} of 31% mismatch power loss is observed for case-1(b) where one-fourth of panels are shaded in these shading conditions.

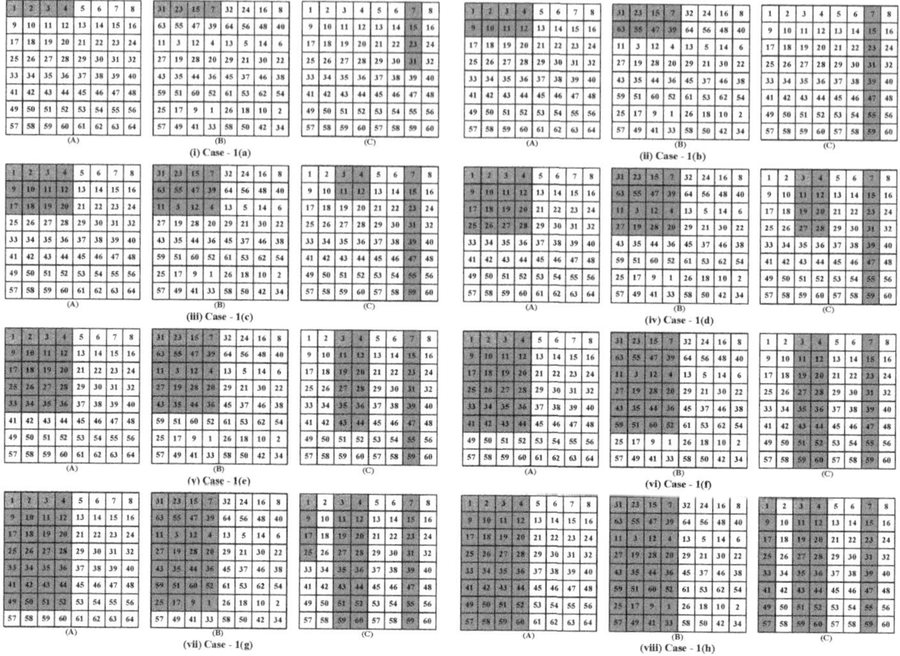

Figure 4.24 Shading pattern for 8×8 SPV Array under progressive incremental Top to Bottom moving shading conditions (A) Shading on conventional configuration (B) Shading on proposed Muddled Baker Map configuration (C) Shade dispersion with proposed configuration.

4.3.2.2 For 6×6 SPV Array Under Progressive Incremental Top to Bottom Moving Shading Conditions

The Solar-Photovoltaic arrangement efficacy over progressive incremental Top to Bottom moving shading scenarios is analyzed using performance metrics in terms of maximum P_O generated and P_{ML} and are explained as follows.

The shading profile under progressive incremental Top to Bottom moving shading conditions for 6×6 SPV array is displayed in Fig. 4.23. The maximum level of output power obtained in the proposed configurations is higher compared against than the conventional configurations which is shown in Table 4.9.

4.3.2.3 For 8×8 SPV Array Under Progressive Incremental Top to Bottom Moving Shading Conditions

The shading pattern for progressive incremental Top to Bottom moving shading conditions along with dispersion of the shade is presented in Fig. 4.24. The power (P_O) generated and the mismatch losses (P_{ML}) for conventional and proposed configuration under each shading case is tabulated in Table 4.10.

Table 4.9

Generated Output Power and Power Loss Caused by Shading for Conventional and Proposed Muddled Baker Map Configuration under Progressive Incremental Top to Bottom Moving Shading Conditions for 6×6 SPV Array

	Topology	Power (W)	Power Loss (W)	Power Loss (%)
Case – 1(a)	SP	2877	1083	27.35
	BL	2812	1148	28.99
	TCT	2816	1144	28.89
	HC	2825	1135	28.66
	Proposed (MBM)	**3245**	**715**	**18.06**
Case – 1(b)	SP	2397	1563	39.47
	BL	2463	1497	37.8
	TCT	2546	1414	35.71
	HC	2475	1485	37.5
	Proposed (MBM)	**2988**	**972**	**24.55**
Case – 1(c)	SP	2385	1575	39.77
	BL	2437	1523	38.46
	TCT	2479	1481	37.4
	HC	2443	1517	38.31
	Proposed (MBM)	**2824**	**1136**	**28.69**
Case – 1(d)	SP	2367	1593	40.23
	BL	2370	1590	40.15
	TCT	2419	1541	38.91
	HC	2370	1590	40.15
	Proposed (MBM)	**2713**	**1247**	**31.49**
Case – 1(e)	SP	2340	1620	40.91
	BL	2346	1614	40.76
	TCT	2362	1598	40.35
	HC	2346	1614	40.76
	Proposed (MBM)	**2416**	**1544**	**38.99**
Case – 1(f)	SP	2309	1651	41.69
	BL	2309	1651	41.69
	TCT	2309	1651	41.69
	HC	2309	1651	41.69
	Proposed (MBM)	**2309**	**1651**	**41.69**

The elaborate study of 4×4, 6×6, and 8×8 Solar-Photovoltaic arrangements reveal that the output energy yield is least in CSP because current levels are constrained due to its series links. The output power (P_O) level developed by the proposed image derived Muddled Baker Map reconfiguration is considerably greater with least mismatch power losses under any type of shading scenarios. The maximum power obtained and its mismatch power loss for progressive incremental diagonal moving shading conditions (Figs. 4.25, 4.27, 4.29) and left to right (Figs. 4.26, 4.28, 4.30) shading conditions are given in Table 4.11 to Table 4.16. From the tables, it can be seen that the proposed method is giving better results than all the existing methods for all the moving and static shading conditions.

Table 4.10

Generated Output Power (W) and Power Loss Caused by Shading for Conventional and Proposed Muddled Baker Map Configuration under Progressive Incremental Top to Bottom Moving Shading Conditions for 8×8 SPV Array

	Topology	Power (W)	Power Loss (W)	Power Loss (%)
Case – 1(a)	SP	5456	836	13.29
	BL	5304	988	15.70
	TCT	5325	967	15.37
	HC	5311	981	15.59
	Proposed (MBM)	**5926**	**366**	**5.82**
Case – 1(b)	SP	4415	1877	29.83
	BL	4606	1686	26.80
	TCT	4482	1810	28.77
	HC	4395	1897	30.15
	Proposed	**5738**	**554**	**8.80**
Case – 1(c)	SP	4259	2033	32.31
	BL	4359	1933	30.72
	TCT	4492	1800	28.61
	HC	4377	1915	30.44
	Proposed (MBM)	**5053**	**1239**	**19.69**
Case – 1(d)	SP	4240	2052	32.61
	BL	4323	1969	31.29
	TCT	4406	1886	29.97
	HC	4332	1960	31.15
	Proposed (MBM)	**4895**	**1397**	**22.20**

Table 4.10
(Continued)

	Topology	Power (W)	Power Loss (W)	Power Loss (%)
Case – 1(e)	SP	4217	2075	32.98
	BL	4233	2059	32.72
	TCT	4326	1966	31.25
	HC	4211	2081	33.07
	Proposed (MBM)	**4763**	**1529**	**24.30**
Case – 1(f)	SP	4185	2107	33.49
	BL	4202	2090	33.22
	TCT	4248	2044	32.49
	HC	4188	2104	33.44
	Proposed (MBM)	**4648**	**1644**	**26.13**
Case – 1(g)	SP	4147	2145	34.09
	BL	4144	2148	34.14
	TCT	4174	2118	33.66
	HC	4152	2140	34.01
	Proposed (MBM)	**4266**	**2026**	**32.20**
Case – 1(h)	SP	4104	2188	34.77
	BL	4104	2188	34.77
	TCT	4104	2188	34.77
	HC	4104	2188	34.77
	Proposed (MBM)	**4104**	**2188**	**34.77**

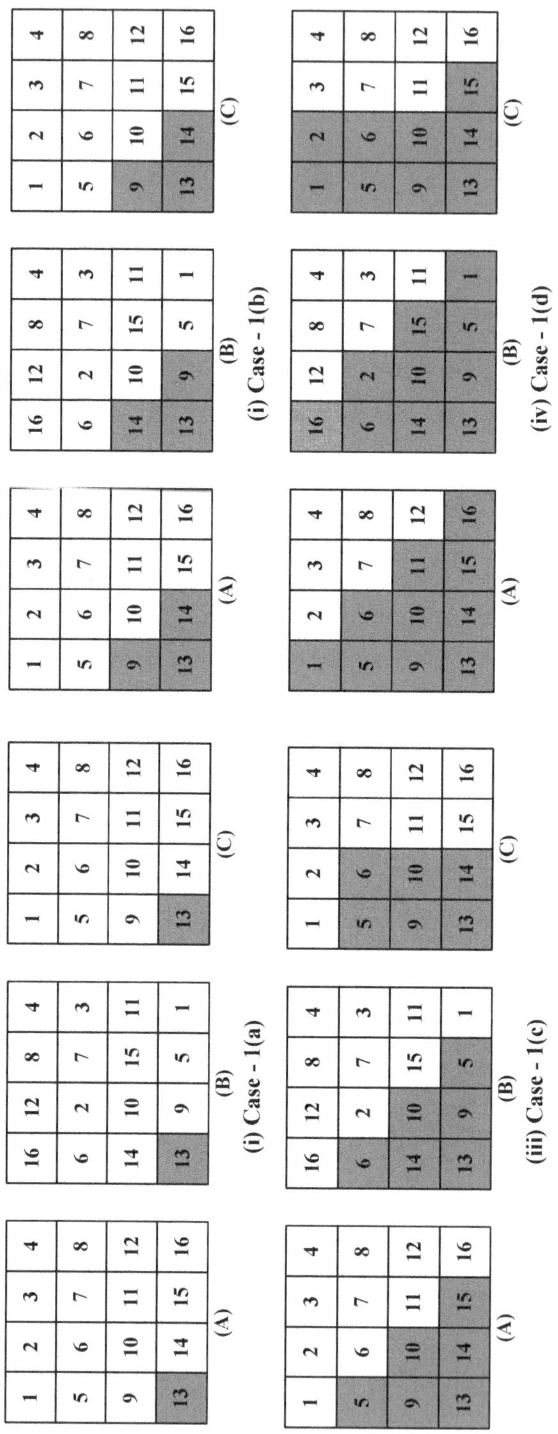

Figure 4.25 Shading pattern for 4×4 SPV array under progressive incremental Diagonal moving shading conditions (A) Shading on Conventional configuration (B) Shading on Proposed Muddled Baker Map configuration (C) Shade dispersion with Proposed configuration.

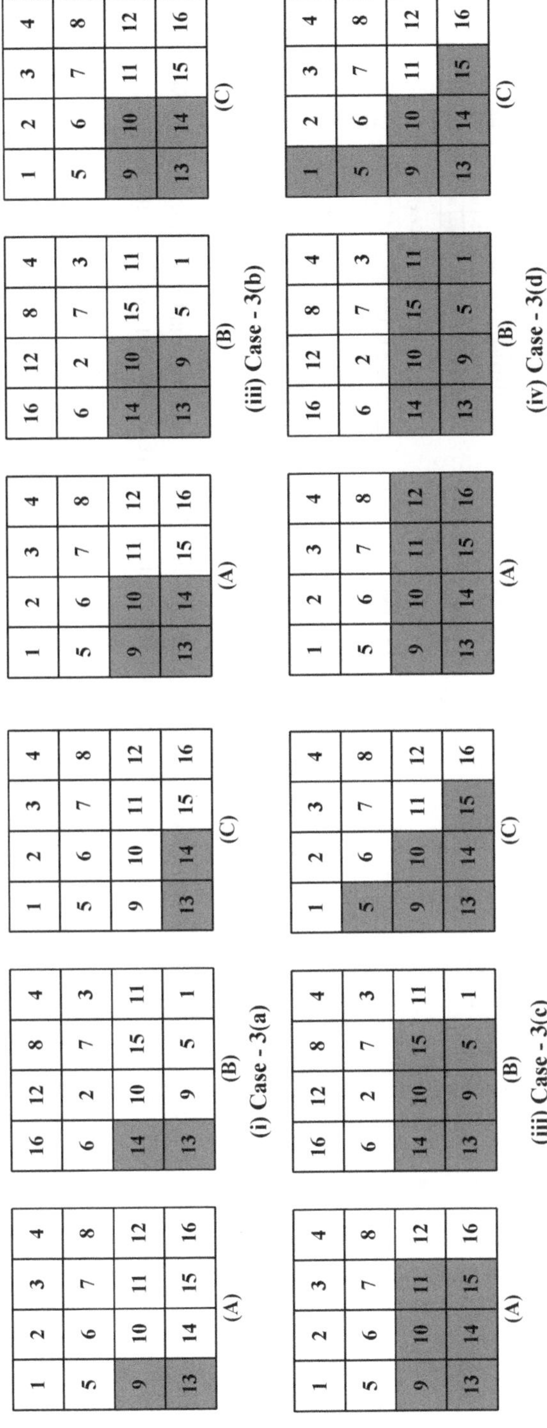

Figure 4.26 Shading pattern for 4×4 SPV array under progressive incremental left to right moving shading conditions (A) Shading on Conventional configuration (B) Shading on proposed Muddled Baker Map configuration (C) Shade dispersion with Proposed configuration.

Photovoltaic Partial Shading

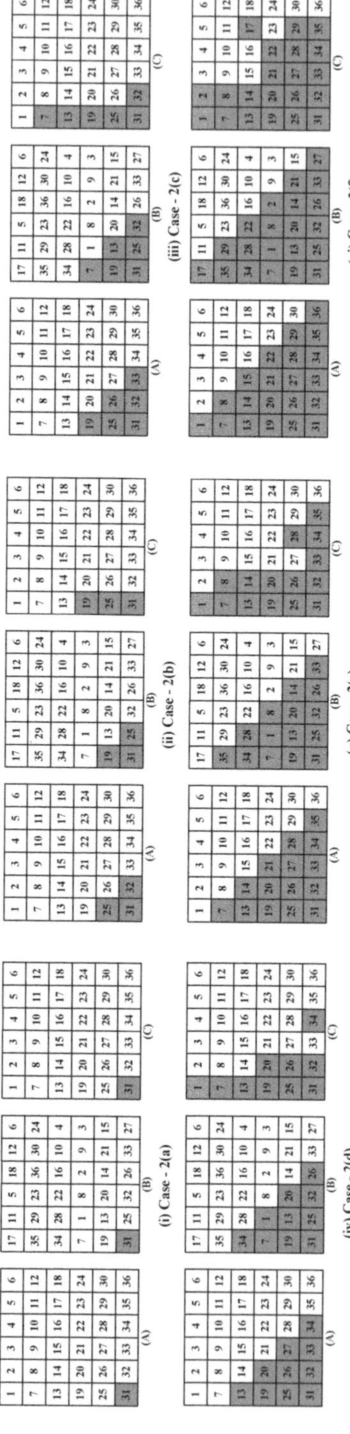

Figure 4.27 Shading pattern for 6 × 6 SPV array under progressive incremental Diagonal moving shading conditions (A) Shading on Conventional configuration (B) Shading on proposed Muddled Baker Map configuration (C) Shade dispersion with Proposed configuration.

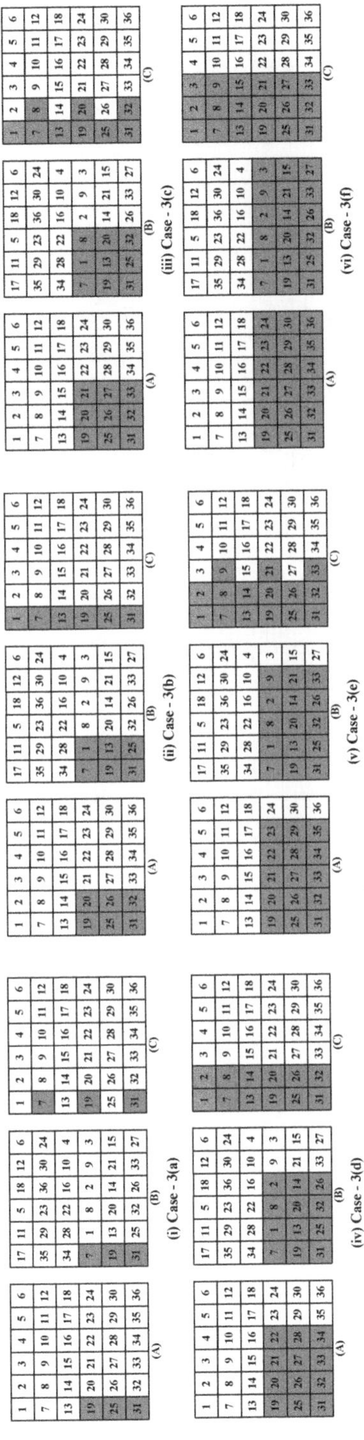

Figure 4.28 Shading pattern for 6×6 SPV array under progressive incremental left to right moving shading conditions (A) Shading on Conventional configuration (B) Shading on proposed Muddled Baker Map configuration (C) Shade dispersion with Proposed configuration.

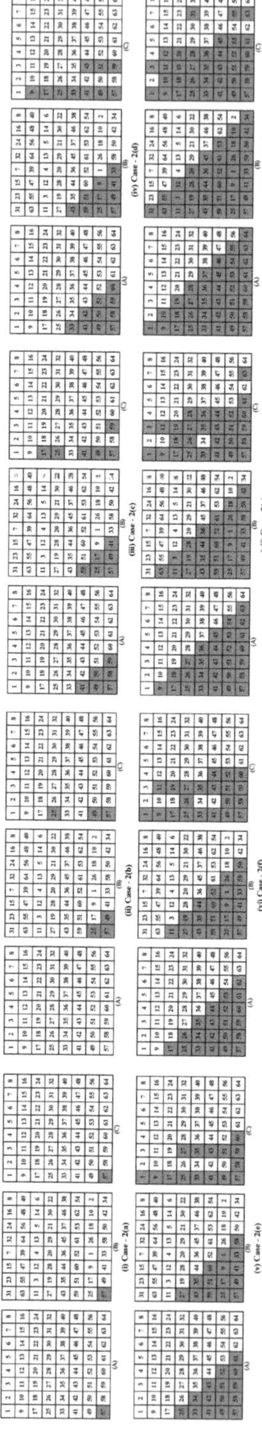

Figure 4.29 Shading pattern for 8×8 SPV array under progressive incremental Diagonal moving shading conditions (A) Shading on Conventional configuration (B) Shading on proposed Muddled Baker Map configuration (C) Shade dispersion with Proposed configuration.

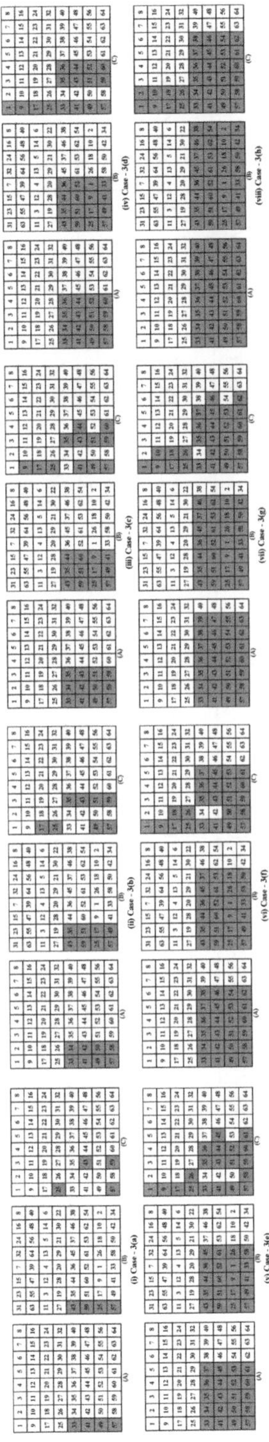

Figure 4.30 Shading pattern for 8×8 SPV array under progressive incremental left to right moving shading conditions (A) Shading on Conventional configuration (B) Shading on proposed Muddled Baker Map configuration (C) Shade dispersion with Proposed configuration.

Table 4.11

Generated Output Power (W) and Power Loss Caused by Shading for Conventional and Proposed Muddled Baker Map Configuration under Progressive Incremental Diagonal Moving Shading Conditions for 4×4 SPV Array

	Maximum Output Power (W)					Power Losses (W)				
	SP	BL	TCT	HC	Proposed (MBM)	SP	BL	TCT	HC	Proposed (MBM)
Case-2(a)	1320	1375	1410	1350	**1410**	253	198	163	223	**163**
Case-2(b)	1063	1099	1139	1117	**1139**	510	474	434	456	**434**
Case-2(c)	798	827	857	836	**1062**	775	746	716	737	**511**
Case-2(d)	551	574	588	584	**706**	1022	999	985	989	**867**

Table 4.12

Generated Output Power (W) and Power Loss Caused by Shading for Conventional and Proposed Muddled Baker Map Configuration under Progressive Incremental Left to Right Moving Shading Conditions for 4×4 SPV Array

	Maximum Output Power (W)					Power Losses (W)				
	SP	BL	TCT	HC	Proposed (MBM)	SP	BL	TCT	HC	Proposed (MBM)
Case-3(a)	1317	1325	1362	1343	**1344**	256	248	211	230	**229**
Case-3(b)	1060	1088	1101	1068	**1101**	513	485	472	505	**472**
Case-3(c)	805	815	828	826	**857**	768	758	745	747	**716**
Case-3(d)	551	551	551	551	**820**	1022	1022	1022	1022	**753**

Table 4.13

Generated Output Power (W) and Power Loss Caused by Shading for Conventional and Proposed Muddled Baker Map Configuration under Progressive Incremental Diagonal Moving Shading Conditions for 6×6 SPV Array

	Maximum Output Power (W)					Power Losses (W)				
	SP	BL	TCT	HC	Proposed (MBM)	SP	BL	TCT	HC	Proposed (MBM)
Case-2(a)	3161	3279	3429	3290	**3429**	378	260	110	249	**110**
Case-2(b)	2780	2882	3030	2892	**3246**	759	658	509	648	**293**
Case-2(c)	2394	2479	2614	2502	**2989**	1145	1060	925	1037	**551**
Case-2(d)	2008	2086	2171	2101	**2784**	1531	1453	1369	1438	**755**
Case-2(e)	1654	1688	1742	1682	**2416**	1886	1852	1797	1858	**1123**
Case-2(f)	1303	1371	1380	1367	**1998**	2236	2168	2160	2173	**1541**

Table 4.14

Generated Output Power and Power Loss Caused by Shading for Conventional and Proposed Muddled Baker Map Configuration under Progressive Incremental Left to Right Moving Shading Conditions for 6×6 SPV Array

	Maximum Output Power (W)					Power Losses (W)				
	SP	BL	TCT	HC	Proposed (MBM)	SP	BL	TCT	HC	Proposed (MBM)
Case-3(a)	3158	3204	3245	3161	**3245**	381	335	294	378	**294**
Case-3(b)	2773	2797	2876	2828	**3127**	766	742	663	711	**412**
Case-3(c)	2601	2660	2707	2640	**3112**	938	879	832	899	**427**
Case-3(d)	2148	2179	2225	2152	**2989**	1391	1360	1314	1387	**550**
Case-3(e)	1735	1732	1733	1732	**1749**	1804	1807	1806	1807	**1790**
Case-3(f)	1290	1290	1290	1290	**2511**	2249	2249	2249	2249	**1028**

Table 4.15

Generated Output Power and Power Loss Caused by Shading for Conventional and Proposed Muddled Baker Map Configuration under Progressive Incremental Diagonal Moving Shading Conditions for 8×8 SPV Array

	Maximum Output Power (W)					Power Losses (W)				
	SP	BL	TCT	HC	Proposed (MBM)	SP	BL	TCT	HC	Proposed (MBM)
Case-2(a)	5860	5958	6157	5971	**6157**	432	334	135	321	**135**
Case-2(b)	5357	5467	5832	5481	**5986**	935	825	460	811	**306**
Case-2(c)	4858	5015	5301	5036	**5785**	1434	1277	991	1256	**507**
Case-2(d)	4350	4466	4712	4522	**5429**	1942	1826	1580	1770	**863**
Case-2(e)	3850	3937	4094	3955	**4675**	2442	2355	2198	2337	**1617**
Case-2(f)	3343	3384	3516	3453	**4462**	2949	2908	2776	2839	**1830**
Case-2(g)	2842	2974	2978	2957	**5023**	3450	3318	3314	3335	**1269**
Case-2(h)	2343	2477	2469	2458	**3337**	3949	3815	3823	3834	**2955**

Table 4.16

Generated Output Power and Power Loss Caused by Shading for Conventional and Proposed Muddled Baker Map Configuration under Progressive Incremental Left to Right Moving Shading Conditions for 8×8 SPV Array

	Maximum Output Power (W)					Power Losses (W)				
	SP	BL	TCT	HC	Proposed (MBM)	SP	BL	TCT	HC	Proposed (MBM)
Case-3(a)	5773	5811	5825	5848	**5926**	519	481	467	444	366
Case-3(b)	5270	5349	5446	5327	**5521**	1022	943	846	965	**771**
Case-3(c)	4752	4796	4940	4771	**5046**	1540	1496	1352	1521	**1246**
Case-3(d)	4240	4323	4403	4332	**4952**	2052	1969	1889	1960	**1340**
Case-3(e)	3735	3788	3871	3802	**4053**	2557	2504	2421	2490	**2239**
Case-3(f)	3222	3293	3314	3257	**3942**	3070	2999	2978	3035	**2350**
Case-3(g)	2711	2747	2761	2759	**3394**	3581	3545	3531	3533	**2898**
Case-3(h)	2204	2204	2204	2204	**4104**	4088	4088	4088	4088	**2188**

4.4 SUMMARY

In this chapter, the reduction of P_{ML} under varied shading situations is suggested using a MBM arrangement reconfiguration method derived from image processing. In the suggested reconfiguration, physically the locations of panels covered in shading are changed to various rows and columns, exclusive of modifying their electric connects. The various uniform and non-uniform shading scenarios were taken to justify the effectiveness of the suggested work. To generalize the framework, the suggested method is analyzed for 4×4, 6×6, and 8×8 Solar-Photovoltaic arrangements. An understanding can be developed from the outcomes that the loss in power is condensed to 22.18%, 15.25%, 51.93%, 25.55% in case of 4×4 Solar-Photovoltaic arrangement, 28.65%, 16.41%, 34.10%, and 23.33% in case of 6×6 Solar-Photovoltaic arrangement and 22.18%, 51.93%, 16.41%, and 23.33% in case of 8×8 Solar-Photovoltaic arrangement with the help of the suggested configuration. The suggested configuration also works well under progressive incremental diagonal moving shading scenarios. The benefit of the suggested configuration is that it does not employ any type of electrical as well as electronic components which includes switches or sensors leading to easy implementation and is also economical in setting up integrated PV (BIPV) and large-sized solar farms under various partial shading situations. The implementation of these methods in practical scenarios is explained in the next chapter along with the experimental setup.

5 Hardware Validation of the Approaches

5.1 INTRODUCTION

In order to test the panels on real-time basis, the approaches discussed in chapter 2 and chapter 3, a lab prototype is developed to test the effect of various progressive incremental shading scenarios on the output power. An experimental setup was developed in order to confirm the efficiency of the approach for previously discussed partial shading conditions as given in Fig. 5.1 to test the solar array on real-time basis. The experimental setup requires various hardware prototypes consisting of solar panels (detailed specifications are given in Table 5.1), connecting wires for joining the panels, ammeter to measure the value of current, voltmeter to read voltage levels, one rheostat, one solar meter to identify the irradiance level, and temperature gun to read the operating temperature of the panels. In order to mimic different irradiance levels, various colors of opaque, transparent, and translucent blocks are employed which blocks the incident light and presents various shading environment. As this is a static configuration-based technique so it does not require the use of sensors and switches and hence there is no requirement of additional electrical and electronics components in the hardware implementation.

The rest of the chapter is organized as follows: Section 5.2 addresses uncertainty analysis of all the measuring devices used for hardware experimentations. Sections 5.3 and 5.4 discusses the practical validation of Tom-Tom and Arrow Sudoku respectively. Finally, Section 5.6 concludes the chapter.

5.2 UNCERTAINTY ANALYSIS OF MEASURING DEVICES USED FOR EXPERIMENTATIONS

For this work, various measuring instruments are employed along with analysis for the uncertainty based on statistical analysis by generating number of readings for these instruments. The process employed for identifying the uncertainty analysis and standard deviations are as follows:

- A range of measurement values is rnoted for identical conditions for a particular device. i.e., $A_1, A_2, \ldots, A_x$, where x is total number measurements.
- The average of the measured values is evaluated using the following equation

$$\overline{A} = \frac{\sum\limits_{i=1}^{x} A_i}{x} \qquad (5.1)$$

DOI: 10.1201/9781003285830-5

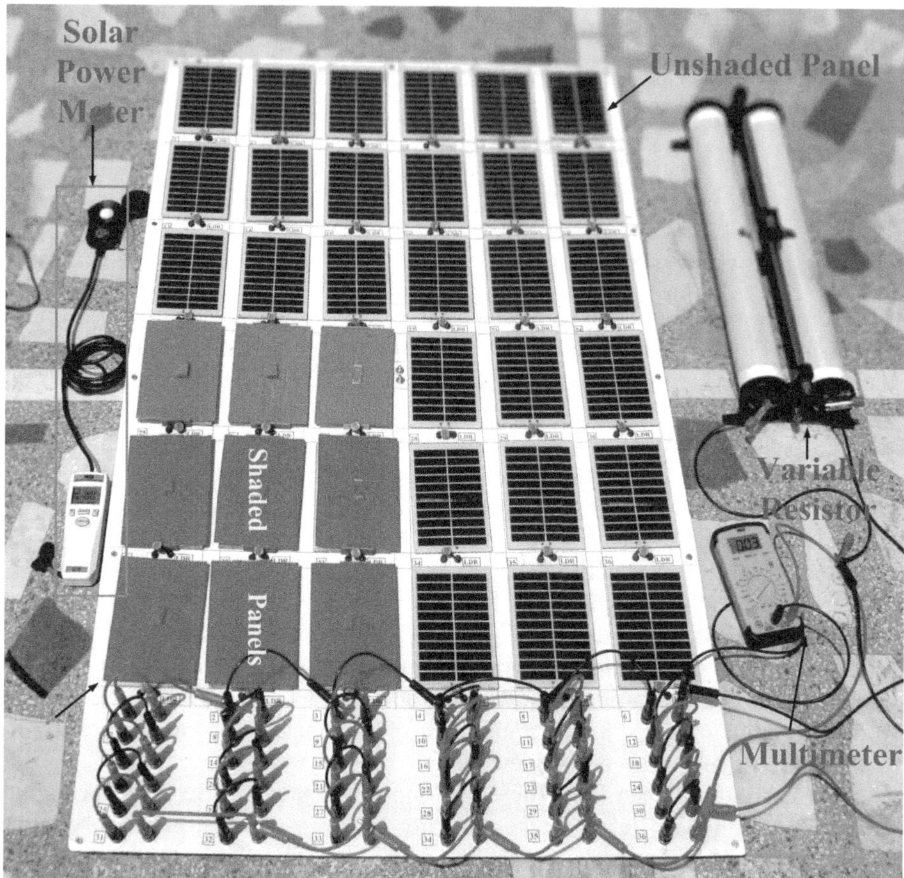

Figure 5.1   Experimental setup of prototype used for validation of the approaches

- The standard deviation (D) for these instruments are calculated by the following equation.

$$D = \sqrt{\frac{\sum\limits_{i=1}^{n}\left(A_i - \overline{A}\right)^2}{(n-1)}} \tag{5.2}$$

5.2.1 MEASUREMENT UNCERTAINTY OF MULTIMETER

The multimeter involved to test the conventional and aforementioned configurations is MECO DMM-63. A Regulated power supply of 20V DC is employed to identify the measurement uncertainty. The multimeter is used to take voltage level measurements and a total five readings are taken, presented in Table 5.2. So, by evaluation,

Table 5.1
Specifications of a Single Panel Used for Experimental Validations

Parameters	Specifications
Panel Power (P)	1.5W
Open Circuit Voltage (V_{oc})	7.15V
Short Circuit Current (I_{sc})	0.27A
Voltage at Maximum Power Point	6V
Current at Maximum Power Point	0.25A

Table 5.2
Values of Voltage Measured by Multimeter for Uncertainty Analysis

Measurement	Value (in Volts)	Deviation From Original Value	Deviation From Average Value
A1	19.99	−0.01	0
A2	20.01	0.01	0.02
A3	20	0	0
A4	20	0	0
A5	19.98	−0.02	−0.01
Average ($\overline{A}$)			19.99
Standard Deviation (D)			0.01224V

the standard deviation for multimeter used in experiment is calculated and the results is $12.24mV$ or $V \pm 0.06\%$.

5.2.2 MEASUREMENT UNCERTAINTY OF SOLAR POWER METER

In order to validate the incident solar irradiation level (G) received by the panels for conventional configurations, a MECO 936 Solar Power Meter is employed. Six different readings are noted at a specific point of time for an irradiation level of $800W/m^2$. The standard deviation is calculated as $G \pm 0.23\%$, which is given in Table 5.3.

5.2.3 MEASUREMENT UNCERTAINTY OF INFRARED TEMPERATURE GUN

The MECO IRT 550P Infrared Thermometer is employed in order to measure the working temperature of the panels. The uncertainty is evaluated by taking 5 readings at $35°C$. The average and standard deviation values are listed in Table 5.4. The standard deviation Standard deviation are evaluated to be $T \pm 0.37\%$.

Table 5.3

Values of Incident Irradiation Measured by Solar Power Meter for Uncertainty Analysis

Measurement	Value (in Volts)	Deviation from	
		Original Value	Average Value
A1	801	1	−1.5
A2	802	2	−2.5
A3	799	−1	0.5
A4	797	−3	2.5
A5	798	−2	1.5
A6	800	0	−0.5
Average ($\overline{A}$)		799.5	
Standard Deviation (D)		$1.8708\,\mathrm{W/m^2}$	

Table 5.4

Values of Operating Temperature Measured by Infrared Thermometer for Uncertainty Analysis

Measurement	Value (in Volts)	Deviation from	
		Original Value	Average Value
A1	34.8	−0.2	0.2
A2	34.9	−0.1	0.1
A3	35.1	0.1	−0.1
A4	35.2	0.2	0.2
A5	35	0	0
Average ($\overline{A}$)		35	
Standard Deviation (D)		$0.13228°C$	

So, from the above analysis it can be concluded that the standard deviations for all measuring instruments is below 1%. Therefore, the measurement devices have negligible impact on system study and can be ignored for further analysis.

5.3 PRACTICAL VALIDATION OF TOM-TOM RECONFIGURATION

The dominance of the Tom-Tom reconfiguration is tested on a 5×5 array and compared with other conventional configurations under progressive incremental shading conditions and different performance metrics such as Maximum output power and

Table 5.5
Specifications of 5×5 Array Used for Experimental Validations

Parameters	Specifications
Panel Power (P)	41.64W
Open Circuit Voltage (V_{oc})	29.92V
Short Circuit Current (I_{sc})	1.39A
Voltage at Maximum Power Point	35.88V
Current at Maximum Power Point	1.5A

Mismatch losses are used to validate its ability. The details of the electrical specifications for a PV array used to validate the Tom-Tom reconfiguration is given in Table 5.5.

The maximum output power produced for SP, BL, TCT, HC, and Tom-Tom configurations experimentally and its mismatch power loss for a 5×5 array under progressive incremental Top to Bottom shading conditions (Fig. 2.14) are shown in Table 5.6. It can be observed from the table that for field test shading conditions, the Tom-Tom method gives enhanced results than the conventional configurations. The output power is enhanced by 19.29% in case-1(a), 23.31% in case-1(b), 20.15% in case-1(c), and 4.66% in case-1(d) in progressive incremental top to bottom shading conditions. There is no increment in power is observed in case-1(e) because shading pattern covers half of the array for each configuration.

The experimental results under progressive incremental diagonal (Fig. 2.16) and left to right shading conditions (Fig. 2.17) are given in Table 5.7 and Table 5.8, respectively. The results confirm show that the maximum power is observed for the configuration than conventional configurations in hardware experimentations also.

5.4 PRACTICAL VALIDATION OF ARROW SUDOKU RECONFIGURATION

The Arrow Sudoku reconfiguration and conventional configurations are tested under progressive incremental shading conditions for 6×6 array and Maximum output power along with Mismatch losses are employed in order to validate the performance of the configurations. The details regarding the electrical specifications of PV array used are given in Table 5.9.

Table 5.10 gives the experimental results of power generated and mismatch loss occurred for SP, BL, TCT, HC, and Arrow Sudoku configurations for 6×6 array under progressive incremental Top to Bottom shading conditions (Fig. 3.14). It is depicted from the table that, under laboratory test shading conditions, the Arrow Sudoku reconfiguration technique yields better performance metrics results than the conventional configurations. The mismatch power loss is reduced to 17.86% in case-1(a), 24.85% in case-1(b), 21.10% in case-1(c), 31.92% in case-1(d), and 37.93% in

Table 5.6

Summary of the Software and Hardware Generated Output Power (W) for Conventional and Tom-Tom Configurations under Progressive Incremental Top to Bottom Shading Conditions for 5×5 SPV Array

	Topology	Software Power (W)	Hardware Power (W)
Case – 1(a)	SP	33.08	30.91
	BL	32.06	29.89
	TCT	31.95	29.78
	HC	32.45	30.28
	(TT)	**39.07**	**36.9**
Case – 1(b)	SP	24.17	22.00
	BL	24.86	22.69
	TCT	25.75	23.58
	HC	25.1	22.93
	(TT)	**32.92**	**30.75**
Case – 1(c)	SP	23.84	21.67
	BL	24.34	22.17
	TCT	24.63	22.46
	HC	24.41	22.24
	(TT)	**30.3**	**28.13**
Case – 1(d)	SP	23.33	21.16
	BL	23.28	21.11
	TCT	23.63	21.46
	HC	23.27	21.1
	(TT)	**24.68**	**22.51**
Case – 1(e)	SP	22.73	20.56
	BL	22.73	20.56
	TCT	22.73	20.56
	HC	22.73	20.56
	(TT)	**22.73**	**20.56**

case-1(e) in progressive incremental top to bottom shading conditions. The final case (case-1(f)) in the Arrow Sudoku reconfigurations gives the same power loss as that of conventional configurations as half of the array is shaded in the vertical direction.

The experimental results under progressive incremental diagonal (Fig. 3.16) and left to right shading conditions (Fig. 3.17) for conventional and Arrow Sudoku reconfigurations are given in Table 5.11 and Table 5.12, respectively. The results clearly show that the maximum power obtained in the configuration is higher than conventional configurations in hardware experimentations also.

Table 5.7

Summary of the Software and Hardware Generated Output Power (W) for Conventional and Tom-Tom Configurations under Progressive Incremental Diagonal Shading Conditions for 5×5 SPV Array

	Software Power (W)					Hardware Power (W)				
	SP	BL	TCT	HC	(TT)	SP	BL	TCT	HC	(TT)
Case-2(a)	37.27	38.46	40.33	37.74	40.33	35.1	36.29	38.16	35.57	**38.16**
Case-2(b)	30.73	32.76	33.84	32.31	38.22	28.56	30.59	31.67	30.14	**36.05**
Case-2(c)	24.43	25.46	26.56	25.33	32.92	22.26	23.29	24.39	23.16	**30.75**
Case-2(d)	18.84	19.45	19.42	19.14	25.12	16.67	17.28	17.25	16.97	**22.95**
Case-2(e)	13.36	13.95	14.01	13.67	22.73	11.19	11.78	11.84	11.5	**20.56**

Table 5.8

Summary of the Software and Hardware Generated Output Power (W) for Conventional and Tom-Tom Configurations under Progressive Incremental Left to Right Shading Conditions for 5×5 SPV Array

	Software Power (W)					Hardware Power (W)				
	SP	BL	TCT	HC	(TT)	SP	BL	TCT	HC	(TT)
Case-3(a)	34.44	34.56	35.29	34.59	**35.29**	32.27	32.39	33.12	32.42	**33.12**
Case-3(b)	29.55	29.98	30.44	30.21	**31.53**	27.38	27.81	28.27	28.04	**29.36**
Case-3(c)	24.66	24.78	25.34	25.15	**29.32**	22.49	22.61	23.17	22.98	**27.15**
Case-3(d)	19.77	20.09	20.12	19.94	**25.3**	17.6	17.92	17.95	17.77	**23.13**
Case-3(e)	14.88	14.88	14.88	14.88	**23.58**	12.71	12.71	12.71	12.71	**21.41**

Table 5.9

Specifications of 6×6 **Array Used for Experimental Validations**

Parameters	Specifications
Panel Power (P)	59.96W
Open Circuit Voltage (V_{oc})	43.06V
Short Circuit Current (I_{sc})	1.80A
Voltage at Maximum Power Point	35.90V
Current at Maximum Power Point	1.67A

5.5 PRACTICAL VALIDATION OF MUDDLED BAKER MAP RECONFIGURATION

To validate the simulated results presented in Chapter 4, Section 4.3.2, the conventional and configurations are tested on 4×4 and 6×6 SPV arrays only. The Muddled Baker Map is compared with all the conventional configurations under progressive incremental shading conditions and the performance is explained below in terms of Maximum output power achieved and Mismatch losses. The detailed electrical specifications of 4×4 and 6×6 SPV arrays used in this work are given in Table 5.13 and Table 5.14, respectively.

Table 5.15 gives the experimental results of power generated and mismatch loss occurred for SP, BL, TCT, HC, and Muddled Baker Map configurations for 4×4 array under progressive incremental Top to Bottom shading conditions (Fig. 4.21). After undergoing shading, the power of 4×4 array is improved to 20.71W in case-1(a), 19.54W in case-1(b) and 16.40W in case-1(c) of progressive incremental top to bottom shading conditions. In the final case, TCT configuration is found to have maximum output power of 15.36W is closely followed by Muddled Baker Map configuration.

The output power and mismatch losses generated under progressive incremental Top to Bottom shading conditions for 6×6 SPV Array under Progressive Incremental Top to Bottom Moving shading conditions (Fig. 4.23) is depicted in Table 5.16. The table reveals that, under field test shading conditions, the Muddled Baker Map method yields a mismatch power loss of 17.86% in case-1(a), 27.19% in case-1(b), 28.58% in case-1(c) and 35.79% in case-1(d). In the case-1(e) and case-1(f), TCT configuration is found to have minimum power loss of 40.73% and 47.47%, respectively and is closely followed by Muddled Baker Map configuration of 45.31% and 47.74%, respectively.

The experimental results under progressive incremental diagonal shading conditions and progressive incremental left to right shading conditions for 4×4 SPV Array is given in Table 5.17 and Table 5.18, respectively and for 6×6 SPV Array, the values are tabulated in Table 5.19 and Table 5.20. The results clearly show that the maximum power obtained in the configuration is higher than conventional configurations in hardware experimentations also.

Table 5.10

Summary of the Software and Hardware Generated Output Power (W) for Conventional and Arrow Sudoku Configurations under Progressive Incremental Top to Bottom Shading Conditions for 6×6 **SPV Array**

	Topology	Software Power (W)	Hardware Power (W)
Case – 1(a)	SP	43.05	40.88
	BL	43.47	41.3
	TCT	45.17	43
	HC	44.12	41.95
	(AS)	**51.42**	**49.25**
Case – 1(b)	SP	38.6	36.43
	BL	39.78	37.61
	TCT	41.32	39.15
	HC	40.01	37.84
	(AS)	**47.23**	**45.06**
Case – 1(c)	SP	38.34	36.17
	BL	39.26	37.09
	TCT	39.97	37.8
	HC	39.37	37.2
	(AS)	**49.48**	**47.31**
Case – 1(d)	SP	37.88	35.71
	BL	37.97	35.8
	TCT	38.78	36.61
	HC	37.97	35.8
	(AS)	**42.99**	**40.82**
Case – 1(e)	SP	37.33	35.16
	BL	37.43	35.26
	TCT	37.71	35.54
	HC	37.45	35.28
	(AS)	**39.39**	**37.22**
Case – 1(f)	SP	36.73	34.56
	BL	36.73	34.56
	TCT	36.73	34.56
	HC	36.73	34.56
	(AS)	**36.73**	**34.56**

Table 5.11

Summary of the Software and Hardware Generated Output Power (W) for Conventional and Arrow Sudoku Configurations under Progressive Incremental Diagonal Shading Conditions for 6×6 SPV Array

	Software Power (W)					Hardware Power (W)				
	SP	BL	TCT	HC	(AS)	SP	BL	TCT	HC	(AS)
Case-2(a)	54.66	56.65	59.19	56.85	**59.19**	53.56	55.55	58.09	55.75	**58.09**
Case-2(b)	48.20	49.92	52.43	50.09	**56.09**	47.10	48.82	51.33	48.99	**54.99**
Case-2(c)	41.65	43.10	45.38	43.49	**51.73**	40.55	42.00	44.28	42.39	**50.63**
Case-2(d)	35.11	36.44	37.87	36.69	**48.27**	34.01	35.34	36.77	35.59	**47.17**
Case-2(e)	29.11	29.69	30.61	29.59	**42.03**	28.01	28.59	29.51	28.49	**40.93**
Case-2(f)	23.17	24.32	24.47	24.25	**34.95**	22.07	23.22	23.37	23.15	**33.85**

Table 5.12

Summary of the Software and Hardware Generated Output Power (W) for Conventional and Arrow Sudoku Configurations under Progressive Incremental Left to Right Shading Conditions for 6×6 SPV Array

	Software Power (W)					Hardware Power (W)				
	SP	BL	TCT	HC	(AS)	SP	BL	TCT	HC	(AS)
Case-3(a)	49.9	50.72	51.42	49.98	**51.42**	47.73	48.55	49.25	47.81	**49.25**
Case-3(b)	44.12	44.57	45.94	45.1	**49.32**	41.95	42.4	43.77	42.93	**47.15**
Case-3(c)	38.34	39.27	39.98	38.99	**44.99**	36.17	37.1	37.81	36.82	**42.82**
Case-3(d)	32.56	33.06	33.73	32.64	**42.99**	30.39	30.89	31.56	30.47	**40.82**
Case-3(e)	26.79	27.38	27.43	27.35	**38.6**	24.62	25.21	25.26	25.18	**36.43**
Case-3(f)	21.03	21.03	21.03	21.03	**36.73**	18.86	18.86	18.86	18.86	**34.56**

Table 5.13

Specifications of 4×4 Array Used for Experimental Validations

Parameters	Specifications
Panel Power (P)	26.65W
Open Circuit Voltage (V_{oc})	28.70V
Short Circuit Current (I_{sc})	1.20A
Voltage at Maximum Power Point	23.93V
Current at Maximum Power Point	1.11A

Table 5.14

Specifications of 6×6 Array Used for Experimental Validations

Parameters	Specifications
Panel Power (P)	59.96W
Open Circuit Voltage (V_{oc})	43.06V
Short Circuit Current (I_{sc})	1.80A
Voltage at Maximum Power Point	35.90V
Current at Maximum Power Point	1.67A

Table 5.15

Summary of the Software and Hardware Generated Output Power (W) for Conventional and Muddled Baker Map under Progressive Incremental Top to Bottom Shading Conditions for 4×4 SPV Array

	Topology	Software Power (W)	Hardware Power (W)
Case – 1(a)	SP	18.4	16.23
	BL	19.07	16.9
	TCT	19.9	17.73
	HC	19.53	17.36
	(MBM)	**22.88**	**20.71**
Case – 1(b)	SP	18.24	16.07
	BL	18.72	16.55
	TCT	18.97	16.8
	HC	18.41	16.24
	(MBM)	**21.71**	**19.54**
Case – 1(c)	SP	17.92	15.75
	BL	17.9	15.73
	TCT	18.19	16.02
	HC	18.05	15.88
	(MBM)	**18.57**	**16.4**
Case – 1(d)	SP	17.53	15.36
	BL	17.53	15.36
	TCT	17.53	15.36
	HC	17.53	15.36
	(MBM)	**14.62**	**12.45**

Table 5.16

Summary of the Software and Hardware Generated Output Power (W) for Conventional and Muddled Baker Map under Progressive Incremental Top to Bottom Shading Conditions for 6×6 SPV Array

	Topology	Software Power (W)	Hardware Power (W)
Case – 1(a)	SP	43.05	40.88
	BL	43.47	41.3
	TCT	45.17	43
	HC	44.12	41.95
	MBM	**51.42**	**49.25**
Case – 1(b)	SP	38.6	36.43
	BL	39.78	37.61
	TCT	41.32	39.15
	HC	40.01	37.84
	MBM	**49.31**	**47.14**
Case – 1(c)	SP	38.34	36.17
	BL	39.26	37.09
	TCT	39.97	37.8
	HC	39.37	37.2
	MBM	**44.99**	**42.82**
Case – 1(d)	SP	37.88	35.71
	BL	37.97	35.8
	TCT	38.78	36.61
	HC	37.97	35.8
	MBM	**40.67**	**38.5**
Case – 1(e)	SP	37.33	35.16
	BL	37.43	35.26
	TCT	37.71	35.54
	HC	37.45	35.28
	MBM	**34.96**	**32.79**
Case – 1(f)	SP	36.73	34.56
	BL	36.73	34.56
	TCT	36.73	34.56
	HC	36.73	34.56
	MBM	**33.5**	**31.33**

Table 5.17

Summary of the Software and Hardware Generated Output Power (W) for Conventional and Muddled Baker Map under Progressive Incremental Diagonal Shading Conditions for 4×4 SPV Array

	Software Power (W)					Hardware Power (W)				
	SP	BL	TCT	HC	(MBM)	SP	BL	TCT	HC	(MBM)
Case-2(a)	22.31	23.15	23.68	22.77	23.68	20.14	20.98	21.51	20.60	**21.51**
Case-2(b)	18.39	18.94	19.55	19.21	19.55	16.22	16.77	17.38	17.04	**17.38**
Case-2(c)	14.35	14.79	15.25	14.93	18.37	12.18	12.62	13.08	12.76	**16.20**
Case-2(d)	10.58	10.93	11.14	11.08	12.94	8.41	8.76	8.97	8.91	**10.77**

Table 5.18

Summary of the Software and Hardware Generated Output Power (W) for Conventional and Muddled Baker Map under Progressive Incremental Left to Right Shading Conditions for 4×4 SPV Array

	Software Power (W)					Hardware Power (W)				
	SP	BL	TCT	HC	(MBM)	SP	BL	TCT	HC	(MBM)
Case-3(a)	22.26	22.39	22.95	22.66	22.68	20.09	20.22	20.78	20.49	**20.51**
Case-3(b)	18.34	18.77	18.97	18.46	18.97	16.17	16.60	16.80	16.29	**16.80**
Case-3(c)	14.45	14.60	14.80	14.77	15.25	12.28	12.43	12.63	12.60	**13.08**
Case-3(d)	10.58	10.58	10.58	10.58	14.68	8.41	8.41	8.41	8.41	**12.51**

Table 5.19

Summary of the Software and Hardware Generated Output Power (W) for Conventional and Muddled Baker Map under Progressive Incremental Diagonal Shading Conditions for 6×6 **SPV Array**

	Software Power (W)					Hardware Power (W)				
	SP	BL	TCT	HC	(MBM)	SP	BL	TCT	HC	(MBM)
Case-2(a)	49.45	51.55	52.86	51.1	**52.86**	47.81	49.91	51.22	49.46	**51.22**
Case-2(b)	43.75	45.56	47.79	46.87	**50.89**	42.11	43.92	46.15	45.23	**49.25**
Case-2(c)	37.99	40.39	41.68	40.51	**46.7**	36.35	38.75	40.04	38.87	**45.06**
Case-2(d)	32.03	33.87	35.15	33.95	**44.46**	30.39	32.23	33.51	32.31	**42.82**
Case-2(e)	26.51	27.31	28.43	27.22	**34.59**	24.87	25.67	26.79	25.58	**32.95**
Case-2(f)	21.18	21.95	22.03	21.95	**27.83**	19.54	20.31	20.39	20.31	**26.19**

Table 5.20

Summary of the Software and Hardware Generated Output Power (W) for Conventional and Muddled Baker Map under Progressive Incremental Left to Right Shading Conditions for 6×6 SPV Array

	Software Power (W)					Hardware Power (W)				
	SP	BL	TCT	HC	(MBM)	SP	BL	TCT	HC	(MBM)
Case-3(a)	49.37	50.19	50.89	49.45	**50.89**	47.73	48.55	49.25	47.81	**49.25**
Case-3(b)	43.59	44.03	45.41	44.57	**48.78**	41.95	42.39	43.77	42.93	**47.14**
Case-3(c)	37.81	38.73	39.44	38.46	**44.46**	36.17	37.09	37.8	36.82	**42.82**
Case-3(d)	32.31	32.52	33.23	32.11	**42.46**	30.67	30.88	31.59	30.47	**40.82**
Case-3(e)	26.26	26.85	26.9	26.81	**38.07**	24.62	25.21	25.26	25.17	**36.43**
Case-3(f)	20.5	20.5	20.5	20.5	**36.2**	18.86	18.86	18.86	18.86	**34.56**

5.6 SUMMARY

In this chapter, the detailed explanation of hardware prototype in field conditions is presented along with the uncertainty analysis. The approaches, Tom-Tom configuration and Arrow Sudoku configuration are tested on progressive incremental left to right, diagonal and top to bottom shading conditions that cover the shading of panel progressively in horizontal, diagonal, and vertical directions. A comprehensive result of values is given for each reconfiguration under these shading conditions. The experiments conducted under this work have clearly demonstrated the outperforming nature of the approaches and the comparison with conventional approaches proves the eminence of the reconfiguration algorithms.

6 Conclusions

In this thesis, efficient methods for reconfiguration of shaded and unshaded panels are developed. The reconfiguration is carried out to disperse the shaded panels throughout the array to mitigate the effects of shading condition. Hence, three approaches for dispersing the shade are developed under this work. These methods were compared in terms of maximum output power and mismatch power loss with the conventional configurations namely SP, BL, TCT, and HC. The performance of these approaches is evaluated on different non-uniform and progressive incremental shading conditions to all shading conditions.

6.1 CONCLUSIONS

The reconfiguration of the panels is employed in order to utilize the ability of the shaded panels so that maximum power can be generated at the output. The static conventional configurations are inefficient in reducing the shading effects. So, to mitigate the effects of shading condition on the panel, a randomized reconfiguration is required. The first approach named as "Tom-Tom puzzle pattern" is to diverse the shading pattern throughout the array. In order to formulate this puzzle pattern, mathematical operations are used as specified in the sub-grids through backtracking. The pattern has to be filled up with the digits 1 to 5 in rows and columns with a constraint of non-repetition. The sub-grids can have same numbers. Each sub-grid points a number and mathematical operand on the left-top corner and the puzzle has to be filled up with the operation required to performed on the numbers for the particular grid.

The second approach named as "Arrow Sudoku" is derived from the family of Sudoku puzzle patterns. The Arrow Sudoku, which is a modified version of original Sudoku, having some similarity to the latter with the difference of the involvement of sub-grids and arrows. The total value after adding the numbers in the squares along the path of each arrow must be equivalent to the number in the circled square. These methods are compared with conventional configuration under different non-uniform shading conditions and progressive incremental moving shading conditions. These reconfigurations outperform many prominent approaches like SP, BL, HC, and TCT. The advantages of these methods are that these reduce the chance of having shaded PV panels in the same row itself and the shade dispersion is dispersed into its own columns.

The third approach named as "Muddled Baker Map" is also to disperse the concentrated shade on to the different rows and columns. In this method, an optimal reconfiguration for Total Cross Tied connection type arrays is to reduce the current difference between its rows. Because of its randomization nature of moving panels to different rows and columns, it is an effective method to relocate the shaded or

DOI: 10.1201/9781003285830-6

Table 6.1
Qualitative Comparison of All Methods

Method ╲ Parameters	P1	P2	P3	P4	P5
Tom-Tom	H	M	L	L	M
Arrow Sudoku	H	M	L	L	M
Muddled Baker Map	H	H	H	H	H

(P_1) Ability to produce maximum power, (P_2) Shade dispersion ability, (P_3) Wiring Complexity, (P_4) Ability to generalize for any size, and (P_5) Income returns.
(L) Low, (M) Moderate, (H) High.

unshaded panels in SPV Array. The reconfiguration strategy is an image processing derived method. Given the short time intervals of partial shading occurrence, the image processing techniques are faster and more cost-effective. The worth of this approach is proved for both non-uniform and progressive incremental moving shading conditions. The approach has provided promising results for both types of shading conditions, when compared with many state-of-the-art approaches.

6.1.1 QUALITATIVE COMPARISON OF ALL APPROACHES

This subsection is dedicated toward all-inclusive comparison of the feature descriptors in this work. The three reconfigurations, namely Tom-Tom, Arrow Sudoku, and Muddled Baker Map, possess different properties, which make them suitable under different partial shading conditions. In order to compare all these approaches, various parameters such as the to generate maximum output power, Ability to disperse the shading pattern under various shading conditions, Wiring Complexity, Ability to generalize so that any array size can be implemented, and income returns are considered. The comparison is shown in Table 6.1.

- All the methods in this book, Arrow Sudoku and Tom-Tom achieve optimal output power under different shading scenarios.
- The methods like Arrow Sudoku and Tom-Tom configuration don't have high dispersion ability because the shade is dispersing in same column only.
- Physical relocation of panels involves a bit of wiring complexity. This depends on the positioning of panels at the optimal place. Within this aspect, they have very low wiring complexity because the panels will be in same columns.
- The recent term, income returns, helps to provide internal understanding to estimate the payback returns before investing it to the array reconfiguration

method. Upon assessment, it can be obtained that Arrow Sudoku and Tom-Tom configuration produces highest returns.
- The main highlight of this comparison is the ability of generalization to any array size. This is not available with Arrow Sudoku and Tom-Tom configurations.

References

1. Anuj Rawat, SK Jha, and Bhavnesh Kumar. Position controlling of sun tracking system using optimization technique. *Energy Reports*, 6:304–309, 2020.

2. Bhavnesh Kumar, SK Jha, and Tarun Kumar. Review of maximum power point tracking techniques for photovoltaic arrays working under uniform/non-uniform insolation level. *International Journal of Renewable Energy Technology*, 9(4):439–452, 2018.

3. Pallavee Bhatnagar and R. K. Nema. Maximum power point tracking control techniques: State-of-the-art in photovoltaic applications. *Renewable and Sustainable Energy Reviews*, 23:224–241, 2013.

4. Omolola A Ogbolumani and Nnamdi Nwulu. Integrated appliance scheduling and optimal sizing of an autonomous hybrid renewable energy system for agricultural food production. In *Advances in Manufacturing Engineering*, pages 651–660. Springer, 2020.

5. Rakesh Reddy Vattigunta, Zakir H Rather, and Ramakrishna Gokaraju. Fast frequency support from hybrid solar PV and wind power plant. In *2018 IEEE International Conference on Power Electronics, Drives and Energy Systems (PEDES)*, pages 1–6. IEEE, 2018.

6. Ujjwal Datta, Akhtar Kalam, and Juan Shi. Battery energy storage system control for mitigating PV penetration impact on primary frequency control and state-of-charge recovery. *IEEE Transactions on Sustainable Energy*, 11(2):746–757, 2019.

7. Tulika Shanker and Ravindra K Singh. Wind energy conversion system: A review. In *2012 Students Conference on Engineering and Systems*, pages 1–6. IEEE, 2012.

8. Rajiv Singh, Asheesh Kumar Singh, and Padmanabh Thakur. Wind Turbine Standards and Certification : Indian Perspective. In *Handbook of Distributed Generation*, chapter 6, pages 205–225. Springer International Publishing, 2017.

9. Abhishek Awasthi, V Karthikeyan, Vipin Das, S Rajasekar, and Asheesh Kumar Singh. Energy Storage Systems in Solar-Wind Hybrid Renewable Systems A Energy management A. In *Smart Energy Grid Design for Island Countries*, chapter 7, pages 189–222. Springer International Publishing, 2017.

10. Manju Aggarwal, Madhusudan Singh, and S K Gupta. Performance Enhancement of Wind Energy System with Distribution Static Compensator. *Indian Journal of Science and Technology*, 10(30):1–8, 2017.

11. VVSN Murty Vallem and Ashwani Kumar. Optimal energy dispatch in microgrids with renewable energy sources and demand response. *International Transactions on Electrical Energy Systems*, 30(5):e12328, 2020.

12. Anongpun Man-Im, Weerakorn Ongsakul, JG Singh, and M Nimal Madhu. Multi-objective optimal power flow considering wind power cost functions using enhanced pso with chaotic mutation and stochastic weights. *Electrical Engineering*, 101(3):699–718, 2019.

13. Peng Wang, Lalit Goel, Xiong Liu, and Fook Hoong Choo. Harmonizing ac and dc: A hybrid ac/dc future grid solution. *IEEE Power and Energy Magazine*, 11(3):76–83, 2013.

14. Pavan Kumar Singh, Nitin Singh, and Richa Negi. Wind Power Forecasting Using Hybrid ARIMA-ANN Technique. In *Ambient Communications and Computer Systems*, pages 209–220. Springer, 2019.

15. Tanuj Rawat, KR Niazi, Nikhil Gupta, and Sachin Sharma. Impact analysis of demand response on optimal allocation of wind and solar based distributed generations in distribution system. *Energy Sources, Part B: Economics, Planning, and Policy*, pages 1–16, 2020.

16. Yingrak Auttawaitkul, Boonnua Pungsiri, Kosin Chammongthai, and Makoto Okuda. A method of appropriate electrical array reconfiguration management for photovoltaic powered car. In *IEEE. APCCAS 1998. 1998 IEEE Asia-Pacific Conference on Circuits and Systems. Microelectronics and Integrating Systems. Proceedings (Cat. No. 98EX242)*, pages 201–204. IEEE, 1998.

17. Atul J Patil, Arush Singh, and RK Jarial. Introduction to condition monitoring of electrical systems. In *Soft Computing in Condition Monitoring and Diagnostics of Electrical and Mechanical Systems*, pages 91–120. Springer, 2020.

18. Ziyad M Salameh and Fouad Dagher. The effect of electrical array reconfiguration on the performance of a PV-powered volumetric water pump. *IEEE Transactions on Energy Conversion*, 5(4):653–658, 1990.

19. Jin Hur Chae-Lim Jeong. A Novel Proposal to Improve Reliability of Spoke-Type BLDC Motor Using Ferrite Permanent Magnet. *IEEE Transactions on Industry Applications*, 52(5):3814–3821, 2016.

20. Damiano La Manna, Vincenzo Li Vigni, Eleonora Riva Sanseverino, Vincenzo Di Dio, and Pietro Romano. Reconfigurable electrical interconnection strategies for photovoltaic arrays: A review. *Renewable and Sustainable Energy Reviews*, 33:412–426, 2014.

21. Rupendra Kumar Pachauri, Om Prakash Mahela, Abhishek Sharma, Jianbo Bai, Yogesh K Chauhan, Baseem Khan, and Hassan Haes Alhelou. Impact of partial shading on various PV array configurations and different modeling approaches: A comprehensive review. *IEEE Access*, 8:181375–181403, 2020.

22. Ibraheem Nasiruddin, Shahida Khatoon, Mohd Faisal Jalil, and R. C. Bansal. Shade diffusion of partial shaded PV array by using odd-even structure. *Solar Energy*, 181(March 2018):519–529, 2019.

23. Zuo Wang, Nanrun Zhou, Lihua Gong, and Minlin Jiang. Quantitative estimation of mismatch losses in photovoltaic arrays under partial shading conditions. *Optik*, 203:163950, 2020.

24. Anurag Singh Yadav, Vinod K Yadav, and Shilpa Choudhary. Power enhancement from solar PV array topologies under partial shading condition. In *2018 International Conference on Power Energy, Environment and Intelligent Control (PEEIC)*, pages 379–383. IEEE, 2018.

25. Moein Jazayeri, Kian Jazayeri, and Sener Uysal. Adaptive photovoltaic array reconfiguration based on real cloud patterns to mitigate effects of non-uniform spatial irradiance profiles. *Solar Energy*, 155:506–516, 2017.

26. Mohammad Amin Ghasemi, Hossein Mohammadian, Mostafa Parniani, and Senior Member. Partial Shading Detection and Smooth Maximum Power Point Tracking of PV Arrays Under PSC. *IEEE Transactions on Power Electronics*, 31(9):6281–6292, 2016.

27. Pankaj Yadav, Amit Kumar, Ankit Gupta, Rupendra Kumar Pachauri, Yogesh K Chauhan, and Vinod Kumar Yadav. Investigations on the effects of partial shading and dust accumulation on PV module performance. In *Proceeding of International Conference on Intelligent Communication, Control and Devices*, pages 1005–1012. Springer, 2017.

28. Santosh Ghosh, Vinod Kumar Yadav, and Vivekananda Mukherjee. Impact of environmental factors on photovoltaic performance and their mitigation strategies–a holistic review. *Renewable Energy Focus*, 28:153–172, 2019.

29. Anurag Singh Yadav and V. Mukherjee. Line losses reduction techniques in puzzled PV array configuration under different shading conditions. *Solar Energy*, 171(March):774–783, 2018.

30. Okan Bingöl and Burcin Özkaya. Analysis and comparison of different PV array configurations under partial shading conditions. *Solar Energy*, 160(December 2017):336–343, 2018.

31. P. Srinivasa Rao, P. Dinesh, G. Saravana Ilango, and C. Nagamani. Optimal Su-Do-Ku based interconnection scheme for increased power output from PV array under partial shading conditions. *Frontiers in Energy*, 9(2):199–210, 2015.

32. Nalin K. Gautam and N.D. Kaushika. An efficient algorithm to simulate the electrical performance of solar photovoltaic arrays. *Energy*, 27(4):347–361, 2002.

33. S Vijayalekshmy S Rama and Iyer Bisharathu. Comparative Analysis on the Performance of a Short String of Series-Connected and Parallel-Connected Photovoltaic Array Under Partial Shading. 2014.

34. Narendra D. Kaushika and Nalin K. Gautam. Energy yield simulations of interconnected solar PV arrays. *IEEE Transactions on Energy Conversion*, 18(1):127–134, 2003.

35. S Malathy and R Ramaprabha. Comprehensive analysis on the role of array size and configuration on energy yield of photovoltaic systems under shaded conditions. *Renewable and Sustainable Energy Reviews*, 49:672–679, 2015.

36. G Velasco, JJ Negroni, F Guinjoan, and R Pique. Energy generation in PV grid-connected systems: power extraction optimization for plant oriented PV generators. In *Proceedings of the IEEE International Symposium on Industrial Electronics, 2005, ISIE 2005.*, volume 3, pages 1025–1030. IEEE, 2005.

37. G Velasco, JJ Negroni, F Guinjoan, and R Pique. Irradiance equalization method for output power optimization in plant oriented grid-connected PV generators. In *2005 European Conference on Power Electronics and Applications*, IEEE, 2005.

38. Guillermo Velasco-Quesada, Francisco Guinjoan-Gispert, Robert Pique-Lopez, Manuel Roman-Lumbreras, and Alfonso Conesa-Roca. Electrical PV array reconfiguration strategy for energy extraction improvement in grid-connected PV systems. *IEEE Transactions on Industrial Electronics*, 56(11):4319–4331, 2009.

39. Guillermo Velasco, Francesc Guinjoan, and Robert Pique. Grid-connected PV systems energy extraction improvement by means of an Electric Array Reconfiguration (EAR) strategy: Operating principle and experimental results. In *2008 IEEE Power Electronics Specialists Conference*, pages 1983–1988. IEEE, 2008.

40. Jonathan P. Storey, Peter R. Wilson, and Darren Bagnall. Improved optimization strategy for irradiance equalization in dynamic photovoltaic arrays. *IEEE Transactions on Power Electronics*, 28(6):2946–2956, 2013.

41. Dzung Nguyen and Brad Lehman. An adaptive solar photovoltaic array using model-based reconfiguration algorithm. *IEEE Transactions on Industrial Electronics*, 55(7):2644–2654, 2008.

42. Dzung D Nguyen, Brad Lehman, and Sagar Kamarthi. Performance evaluation of solar photovoltaic arrays including shadow effects using neural network. In *2009 IEEE Energy Conversion Congress and Exposition*, pages 3357–3362. IEEE, 2009.

43. Shubhankar Niranjan Deshkar, Sumedh Bhaskar Dhale, Jishnu Shekar Mukherjee, T. Sudhakar Babu, and N. Rajasekar. Solar PV array reconfiguration under partial shading conditions for maximum power extraction using genetic algorithm. *Renewable and Sustainable Energy Reviews*, 43(2014):102–110, 2015.

44. Naik Aashay Rajan, Kulkarni Devendra Shrikant, B. Dhanalakshmi, and N. Rajasekar. Solar PV array reconfiguration using the concept of Standard deviation and Genetic Algorithm. *Energy Procedia*, 117:1062–1069, 2017.

45. Hongda Liu, Lidong Wan, and Biying Pei. Optimal scheme of PV array in partial shading: A reconfiguration algorithm. In *2016 IEEE International Conference on Mechatronics and Automation*, pages 2105–2109. IEEE, 2016.

46. N Gupta, A Swarnkar, KR Niazi, and RC Bansal. Multi-objective reconfiguration of distribution systems using adaptive genetic algorithm in fuzzy framework. *IET generation, transmission & distribution*, 4(12):1288–1298, 2010.

47. Yamille Del Valle, Ganesh Kumar Venayagamoorthy, Salman Mohagheghi, Jean-Carlos Hernandez, and Ronald G Harley. Particle swarm optimization: basic concepts, variants and applications in power systems. *IEEE Transactions on evolutionary computation*, 12(2):171–195, 2008.

48. Raghavendra V Kulkarni and Ganesh Kumar Venayagamoorthy. Particle swarm optimization in wireless-sensor networks: A brief survey. *IEEE Transactions on Systems, Man, and Cybernetics, Part C (Applications and Reviews)*, 41(2):262–267, 2010.

49. Haluk Gozde and M Cengiz Taplamacioglu. Automatic generation control application with craziness based particle swarm optimization in a thermal power system. *International Journal of Electrical Power & Energy Systems*, 33(1):8–16, 2011.

50. Sankhadeep Chatterjee, Sarbartha Sarkar, Sirshendu Hore, Nilanjan Dey, Amira S Ashour, and Valentina E Balas. Particle swarm optimization trained neural network for structural failure prediction of multistoried RC buildings. *Neural Computing and Applications*, 28(8):2005–2016, 2017.

51. Natwar S Rathore, VP Singh, and Bui Duc Hong Phuc. A modified controller design based on symbiotic organisms search optimization for desalination system. *Journal of Water Supply: Research and Technology-Aqua*, 68(5):337–345, 2019.

52. T. Sudhakar Babu, J. Prasanth Ram, Tomislav Dragicevic, Masafumi Miyatake, Frede Blaabjerg, and Natarajan Rajasekar. Particle Swarm Optimization based Solar PV Array Reconfiguration of the Maximum Power Extraction under Partial Shading Conditions. *IEEE Transactions on Sustainable Energy*, 9(1):74–85, 2017.

53. Haluk Gozde, M Cengiz Taplamacioglu, and Ilhan Kocaarslan. Comparative performance analysis of artificial bee colony algorithm in automatic generation control for interconnected reheat thermal power system. *International Journal of Electrical Power & Energy Systems*, 42(1):167–178, 2012.

54. Mohammadreza Akrami and Kazem Pourhossein. A novel reconfiguration procedure to extract maximum power from partially-shaded photovoltaic arrays. *Solar Energy*, 173(December 2017):110–119, 2018.

55. Dalia Yousri, Sudhakar Babu Thanikanti, Karthik Balasubramanian, Ahmed Osama, and Ahmed Fathy. Multi-objective grey wolf optimizer for optimal design of switching matrix for shaded PV array dynamic reconfiguration. *IEEE Access*, 8:159931–159946, 2020.

56. Manjunath Matam, Venugopal Reddy Barry, and Avinash Reddy Govind. Optimized reconfigurable PV array based photovoltaic water-pumping system. *Solar Energy*, 170:1063–1073, 2018.

57. Koray Sener Parlak. PV array reconfiguration method under partial shading conditions. *International Journal of Electrical Power and Energy Systems*, 63:713–721, 2014.

58. Koray Sener Parlak and Mehmet Karakose. An efficient reconfiguration method based on standard deviation for series and parallel connected PV arrays. In *3rd International Conference on Renewable Energy Research and Applications, ICRERA 2014*, pages 457–461, 2014.

59. D Picault, B Raison, S Bacha, J De Casa, and J Aguilera. Forecasting photovoltaic array power production subject to mismatch losses. *Solar Energy*, 84(7):1301–1309, 2010.

60. Luiz Fernando Lavado Villa, Damien Picault, Bertrand Raison, Seddik Bacha, and Antoine Labonne. Maximizing the power output of partially shaded photovoltaic plants through optimization of the interconnections among its modules. *IEEE Journal of Photovoltaics*, 2(2):154–163, 2012.

61. Chayut Tubniyom, Watcharin Jaideaw, Rongrit Chatthaworn, Amnart Suksri, and Tanakorn Wongwuttanasatian. Effect of partial shading patterns and degrees of shading on Total Cross-Tied (TCT) photovoltaic array configuration. *Energy Procedia*, 153:35–41, 2018.

62. Chayut Tubniyom, Rongrit Chatthaworn, Amnart Suksri, and Tanakorn Wongwuttanasatian. Minimization of Losses in Solar Photovoltaic Modules by Reconfiguration under Various Patterns. *Energies*, 12(24), 2019.

63. Kuei Hsiang Chao, Pei Lun Lai, and Bo Jyun Liao. The optimal configuration of photovoltaic module arrays based on adaptive switching controls. *Energy Conversion and Management*, 100:157–167, 2015.

64. B. Indu Rani, G. Saravana Ilango, and Chilakapati Nagamani. Enhanced power generation from PV array under partial shading conditions by shade dispersion using Su Do Ku configuration. *IEEE Transactions on Sustainable Energy*, 4(3):594–601, 2013.

65. Srinivasa Rao Potnuru, Dinesh Pattabiraman, Saravana Ilango Ganesan, and Nagamani Chilakapati. Positioning of PV panels for reduction in line losses and mismatch losses in PV array. *Renewable Energy*, 78:264–275, 2015.

66. G. Sai Krishna and Tukaram Moger. Improved SuDoKu reconfiguration technique for total-cross-tied PV array to enhance maximum power under partial shading conditions. *Renewable and Sustainable Energy Reviews*, 109(December 2018):333–348, 2019.

67. Majid Horoufiany and Reza Ghandehari. Optimization of the Sudoku based reconfiguration technique for PV arrays power enhancement under mutual shading conditions. *Solar Energy*, 159:1037–1046, 2018.

68. Majid Horoufiany and Reza Ghandehari. Optimization of the sudoku based reconfiguration technique for PV arrays power enhancement under mutual shading conditions. *Solar Energy*, 159:1037–1046, 2018.

69. S Vijayalekshmy, GR Bindu, and S Rama Iyer. Analysis of various photovoltaic array configurations under shade dispersion by su do ku arrangement during passing cloud conditions. *Indian Journal of Science and Technology*, 8(35), 2015.

70. S. Vijayalekshmy, G. R. Bindu, and S. Rama Iyer. A novel Zig-Zag scheme for power enhancement of partially shaded solar arrays. *Solar Energy*, 135:92–102, 2016.

71. Warisha Meraj and Sukumar Mishra. An approach to increase power in partial shading condition by reconfiguration. In *2015 Annual IEEE India Conference (INDICON)*, pages 1–6. IEEE, 2015.

72. Anurag Singh Yadav, Rupendra Kumar Pachauri, and Yogesh K. Chauhan. Comprehensive investigation of PV arrays with puzzle shade dispersion for improved performance. *Solar Energy*, 129:256–285, 2016.

73. Anurag Singh Yadav, Rupendra Kumar Pachauri, Yogesh K. Chauhan, S. Choudhury, and Rajesh Singh. Performance enhancement of partially shaded PV array using novel shade dispersion effect on magic-square puzzle configuration. *Solar Energy*, 144:780–797, 2017.

74. Namani Rakesh and T. Venkata Madhavaram. Performance enhancement of partially shaded solar PV array using novel shade dispersion technique. *Frontiers in Energy*, 10(2):227–239, 2016.

75. Namani Rakesh, T Venkata Madhavaram, K Ajith, G Rajendra Naik, and P Nagarjun Reddy. A new technique to enhance output power from solar PV array under different partial shaded conditions. In *2015 IEEE International Conference on Electron Devices and Solid-State Circuits (EDSSC)*, pages 345–348. IEEE, 2015.

76. Sarojini Mary Samikannu, Rakesh Namani, and Senthil Kumar Subramaniam. Power enhancement of partially shaded PV arrays through shade dispersion using magic square configuration. *Journal of Renewable and Sustainable Energy*, 8(6), 2016.

77. Manjunath, H. N. Suresh, and S. Rajanna. Performance enhancement of Hybrid interconnected Solar Photovoltaic array using shade dispersion Magic Square Puzzle Pattern technique under partial shading conditions. *Solar Energy*, 194(May):602–617, 2019.

78. Lahcen El Iysaouy, Mhammed Lahbabi, and Abdelmajid Oumnad. A novel magic square view topology of a PV system under partial shading condition. *Energy Procedia*, 157(2018):1182–1190, 2019.

79. S. Malathy and R. Ramaprabha. Reconfiguration strategies to extract maximum power from photovoltaic array under partially shaded conditions. *Renewable and Sustainable Energy Reviews*, (April):1–13, 2017.

80. M. John Bosco and M. Carolin Mabel. A novel cross diagonal view configuration of a PV system under partial shading condition. *Solar Energy*, 158(November 2016):760–773, 2017.

81. Reddy Sreekantha and Yammani Chandrasekhar. A novel Magic-Square puzzle based one-time PV reconfiguration technique to mitigate mismatch power loss under various partial shading conditions. *Optik*, 2020.

82. Priya Ranjan Satpathy, Renu Sharma, and Sasmita Jena. A shade dispersion interconnection scheme for partially shaded modules in a solar PV array network. *Energy*, 139:350–365, 2017.

83. Priya Ranjan Satpathy and Renu Sharma. Power loss reduction in partially shaded PV arrays by a static SDP technique. *Energy*, 156:569–585, 2018.

84. N. Belhaouas, M.-S. Ait Cheikh, P. Agathoklis, M.-R. Oularbi, B. Amrouche, K. Sedraoui, and N. Djilali. PV array power output maximization under partial shading using new shifted PV array arrangements. *Applied Energy*, 187:326–337, 2017.

85. Gurusamy Madhusudanan, Subramaniam Senthilkumar, I. Anand, and Padmanaban Sanjeevikumar. A shade dispersion scheme using Latin square arrangement to enhance power production in solar photovoltaic array under partial shading conditions. *Journal of Renewable and Sustainable Energy*, 10(5):053506, 2018.

86. S Sreekantha Reddy and Chandrasekhar Yammani. Odd-Even-Prime pattern for PV array to increase power output under partial shading conditions. *Energy*, 213:118780, 2020.

87. Himanshu Sekhar Sahu, Sisir Kumar Nayak, and Sukumar Mishra. Maximizing the Power Generation of a Partially Shaded PV Array. *IEEE Journal of Emerging and Selected Topics in Power Electronics*, 4(2):626–637, 2016.

88. B. Dhanalakshmi and N. Rajasekar. A novel Competence Square based PV array reconfiguration technique for solar PV maximum power extraction. *Energy Conversion and Management*, 174(June):897–912, 2018.

89. B. Dhanalakshmi and N. Rajasekar. Dominance square based array reconfiguration scheme for power loss reduction in solar PhotoVoltaic (PV) systems. *Energy Conversion and Management*, 156(September 2017):84–102, 2018.

90. Dhanup S. Pillai, N. Rajasekar, J. Prasanth Ram, and Venkatachalam Kumar Chinnaiyan. Design and testing of two phase array reconfiguration procedure for maximizing power in solar PV systems under partial shade conditions (PSC). *Energy Conversion and Management*, 178(June):92–110, 2018.

91. Rupendra Pachauri, Anurag Singh Yadav, Yogesh K Chauhan, Abhinav Sharma, and Vinod Kumar. Shade dispersion-based photovoltaic array configurations for performance enhancement under partial shading conditions. *International Transactions on Electrical Energy Systems*, 28(7):e2556, 2018.

92. Neha Mishra, Anurag Singh Yadav, Rupendra Pachauri, Yogesh K. Chauhan, and Vinod K. Yadav. Performance enhancement of PV system using proposed array topologies under various shadow patterns. *Solar Energy*, 157:641–656, 2017.

93. Priya Ranjan Satpathy and Renu Sharma. Power and mismatch losses mitigation by a fixed electrical reconfiguration technique for partially shaded photovoltaic arrays. *Energy Conversion and Management*, 192(January):52–70, 2019.

94. Dhanup S. Pillai, J. Prasanth Ram, Malisetty Siva Sai Nihanth, and N. Rajasekar. A simple, sensorless and fixed reconfiguration scheme for maximum power enhancement in PV systems. *Energy Conversion and Management*, 172(May):402–417, 2018.

95. Malisetty Siva Sai Nihanth, J. Prasanth Ram, Dhanup S. Pillai, Amer M.Y.M. Ghias, Akhil Garg, and N. Rajasekar. Enhanced power production in PV arrays using a new skyscraper puzzle based one-time reconfiguration procedure under partial shade conditions (PSCs). *Solar Energy*, 194(May):209–224, 2019.

96. Namani Rakesh, S. Senthil Kumar, and G. Madhusudanan. Mitigation of power mismatch losses and wiring line losses of partially shaded solar PV array using improvised magic technique. *IET Renewable Power Generation*, 13(9):1522–1532, 2019.

97. G Madhusudanan, N Rakesh, S Senthil Kumar, and S Sarojini Mary. Solar photovoltaic array reconfiguration using magic su-do-ku algorithm for maximum power production under partial shading conditions. *International Journal of Ambient Energy*, pages 1–12, 2019.

98. Priya Ranjan Satpathy, Renu Sharma, and Sambit Dash. An efficient SD-PAR technique for maximum power generation from modules of partially shaded PV arrays. *Energy*, 175:182–194, 2019.

99. G. Meerimatha and B. Loveswara Rao. Novel reconfiguration approach to reduce line losses of the photovoltaic array under various shading conditions. *Energy*, 196:117120, 2020.

100. Ensherah A. Naeem, Mustafa M. Abd Elnaby, Hala S. El-sayed, Fathi E. Abd El-Samie, and Osama S. Faragallah. Wavelet fusion for encrypting images with a few details. *Computers and Electrical Engineering*, 54:450–470, 2016.

101. Zhengjun Liu, Yu Zhang, Wei Liu, Fanyi Meng, Qun Wu, and Shutian Liu. A mixed scrambling operation for hiding image. *Optik*, 124(22):5391–5396, 2013.

102. Ensherah A. Naeem, Mustafa M. Abd Elnaby, Naglaa F. Soliman, Alaa M. Abbas, Osama S. Faragallah, Noura Semary, Mohiy M. Hadhoud, Saleh A. Alshebeili, and Fathi E. Abd El-Samie. Efficient implementation of chaotic image encryption in transform domains. *Journal of Systems and Software*, 97:118–127, 2014.

103. Zhengjun Liu, She Li, Wei Liu, and Shutian Liu. Opto-digital image encryption by using Baker mapping and 1-D fractional Fourier transform. *Optics and Lasers in Engineering*, 51(3):224–229, 2013.

104. Gonzalo Alvarez and Shujun Li. Breaking an encryption scheme based on chaotic baker map. *Physics Letters, Section A: General, Atomic and Solid State Physics*, 352(1-2):78–82, 2006.

105. Jiri Fridrich. Symmetric Ciphers Based on Two-Dimensional Chaotic Maps. *International Journal of Bifurcation and Chaos*, 8(6):1259–1284, 1998.

106. K.A. Abitha and Pradeep K. Bharathan. Secure Communication Based on Rubik's Cube Algorithm and Chaotic Baker Map. In *Procedia Technology*, volume 24, pages 782–789. Elsevier B.V., 2016.

Index